W9-BPM-266

The
# Astronomy
# Bible

# The
# Astronomy
# Bible

**The definitive guide to the
night sky and the universe**

**Heather Couper
& Nigel Henbest**

# FIREFLY BOOKS

# A FIREFLY BOOK

Published by Firefly Books Ltd. 2015

First printing

**Publisher Cataloging-in-Publication Data (U.S.)**

A CIP record for this title is available from the Library of Congress

**Library and Archives Canada Cataloguing in Publication**

Couper, Heather
    The Astronomy Bible : the definitive guide to the night sky and the universe / Heather Couper and Nigel Henbest
Includes index
ISBN 978-1-77085-482-6 (pbk.)
    1. Astronomy--Popular works.  I. Henbest, Nigel, author  II. Title.
QB44.3.C678 2015 520 C2014-906113-7:

Published in the United States by
Firefly Books (U.S.) Inc.
P.O. Box 1338, Ellicott Station
Buffalo, New York 14205

Published in Canada by
Firefly Books Ltd.
50 Staples Avenue, Unit 1
Richmond Hill, Ontario L4B 0A7

Printed and bound in China

First published in Great Britain
in 2015 by Philip's, a division of
Octopus Publishing Group Ltd,
Endeavour House, 189 Shaftesbury
Avenue, London WC2H 8JY

# Contents

Introduction                                      6

Observing the sky                               22
The Moon                                         64
Planets                                          88
Cosmic vermin                                  142
Sun                                             178
Stars                                           202
Cosmos                                          250
Constellations                                 276
Reference                                       370

Index                                           395
Acknowledgments                                399
Photo credits                                   400

Throughout this book, distances are given in kilometers. To convert kilometers to miles, simply divide by 1.609. To convert miles to kilometers, multiply by the same figure. Focal lengths of binoculars are generally given in millimeters. Apertures of telescopes are quoted in centimeters, although dealers will often give them in inches; to convert inches to centimeters, multiply by 2.54; to make the reverse conversion, divide by that figure. There are, of course, 10 millimeters to a centimeter.

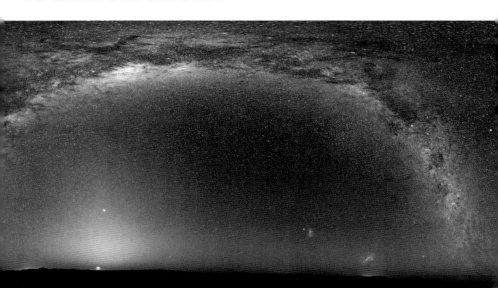

# CHAPTER 1
# INTRODUCTION

# WHAT IS ASTRONOMY?

Picture a breathtakingly clear starry, starry night. The heavens are spangled with diamonds, glittering against the black velvet of space…

Each of these thousands of stars has a personality; each has its own story to tell.

Our ancestors saw the sky as a starscape, just as we see a landscape on Earth. And, just as we make maps of our world, they joined up the dots in the sky, making them into the constellation patterns we know today. To ensure that farmers and navigators at sea knew the star patterns, they invoked well-known stories to describe them.

Against this tableau of bears, giants, celestial crosses and flying swans, the heavens can give us surprises: an outburst of shooting stars; a comet; the glorious display of the Northern or Southern lights, with their swirling curtains of red and green, or even an exploding star.

Then we have our much-loved Moon, whose face changes from day-to day as it

orbits the Earth. And the planets, from Mercury to Neptune, grace our skies every night of the year. Our local star, the Sun, puts in an appearance every day (weather permitting).

This is astronomy: a celebration of the beauty of the heavens, and a science that now brings in physics, chemistry, mathematics and biology.

But the glory of astronomy is that it's not brain surgery. Anyone can do it.

You can delight in the heavens by lying on your steamer or deck chair and gazing at the stars and planets, or you can bring them closer by using binoculars or a telescope. You could even go the whole way and do an astronomy or astrophysics degree at university. That is what you need to be a pro.

Professional astronomy today is truly awesome. A whole armory of sophisticated and powerful equipment around the world is trained on the sky, as well as massive telescopes in space. Who can forget the "Pillars of Creation" image from the Hubble Space Telescope?

Astronomers launch probes to the planets, and one of them may have found primitive life on Mars. They have discovered objects beyond the imagination of their predecessors: supernovae (exploding stars), white dwarfs, quasars, pulsars (ultra-dense stars whirling around faster than you can blink) and black holes. They have even pinned down the origin of the Universe itself to a Big Bang nearly 14 billion years ago.

Astronomy is an adventure and this book shows how you can be part of it. The sky's the limit!

*A sight to inspire: a stunningly-clear sky, with the Milky Way arching overhead.*

# THE ROOTS OF ASTRONOMY

Astronomy's history goes back at least 40,000 years, to the Aboriginal peoples under the coal-black heavens of Australia. With no light pollution they were crowded with a blanket of brilliant stars, so the native Australians devised patterns in the sky that centered around the dark patches in the heavens (which we now know to be sooty cosmic clouds poised to form new stars). They even invented a dark constellation called the Emu.

Observers in Babylon later started to systematize the sky, joining up the stellar dots into constellations that reflected the stories of their mythology. Ancient constellations like Leo, the lion, have been with us for some 4,000 years.

Early astronomers realized that the night sky was not just a beautiful panorama, but that it was also a practical tool. You could use the patterns in the sky for timekeeping, as they moved regularly across the sky every night. You could also

*A Polynesian voyaging canoe, steering by the stars*

*The mighty megaliths of Stonehenge, in England were aligned to keep track of the seasons.*

utilize them in navigation at sea. Indeed, Greek and the Polynesian sailors both set their sights on the stars. The heavens became essential for calendar-making; our ancestors observed that you could see different constellations each month during the year.

They also observed the movements of the Sun and Moon. Early civilizations marked the passing of the year by watching the changing position of sunrise and sunset along the horizon (as you can, as well).

They built vast stone monuments to our local star, the most famous of which is Stonehenge, in Wiltshire, England, constructed around 2500 BC. Traditionally, visitors to the monument celebrate at the summer solstice, normally June 21, when from inside the circle the rising Sun aligns with the outlying "heel stone" to mark the height of summer. But a more recent theory is that Stonehenge is a monument dedicated to winter, and that an observer would stand at the heel stone and watch the Sun setting between two towering sarsen stones on winter's shortest day. It marked the turning point of the year, and the coming of spring.

But astronomy was still not a science. For this, we would have to wait over a thousand years.

# MAKING SENSE OF THE HEAVENS

The Greeks, with their logical and analytical minds, were the first civilization to turn astronomy into a science. Confusingly, most of the highly gifted astronomer/philosophers didn't live on the Greek mainland, but in scattered parts of its huge empire: Sicily, Turkey and Egypt's Alexandria.

They observed eclipses of the Moon, and concluded that these were caused by our companion moving into the Earth's shadow. From the curve of the shadow on the Moon, they deduced that the Earth was spherical, and about four times larger than our satellite.

In around 240 BC, the polymath Eratosthenes accurately calculated the circumference of the Earth, by comparing the overhead positions of the midsummer Sun at two different locations in Egypt.

Far from being abstract philosophers, the Greeks invented the first computer: the Antikythera mechanism. With its complex set of gearwheels, it was used to predict the Moon's phases and eclipses.

The Greek astronomers also watched the planets (which we now know to be neighboring worlds) — "wandering stars" that moved from night to night, and they worked out a template for our Solar System.

## PTOLEMY'S UNIVERSE

Ptolemy and most of his colleagues believed that the planets circled the Earth. So why did some (especially Mars) loop backwards and forwards in the sky? Ptolemy's answer was logical: a planet goes around in its own big circle, but it rides on a little "wheel" (epicycle) attached to the main orbit, looping around itself and criss-crossing our heavens as a result.

A corroded gearwheel, found in a shipwreck off the Greek island of Antikythera, is the largest remaining fragment of a 2,000-year-old computer that predicted eclipses and the Moon's phases.

The architect of this grand plan was the last great Greek astronomer, Alexandria's Claudius Ptolemy. He collated all the teachings of his predecessors into an astronomy bible called the *Almagest*, which was in publication for 1,500 years. It also ensured that the Greek findings tumbled down through the centuries. Ptolemy was the first person to catalogue the heavens systematically. He listed 1,022 stars, grouped into 48 constellations.

However, with the end of the Greek Empire, astronomy lapsed into the doldrums. It would be another 1,000 years before it was resuscitated; this time, in the Middle East. Persian astronomers gave us the names to some of our most famous stars, such as the variable star Albebaran ("the follower," because it stalks the Pleiades). But after more than a millennium, astronomy was in need of a breakthrough.

# REVOLUTION IN THE COSMOS

In 1543, Nicolaus Copernicus created a revolution in astronomy, in more ways than one. The administrator of Frombork cathedral in north-eastern Poland, Copernicus observed the movements of the planets from a large tower in the grounds, and realized that their positions were well away from where Ptolemy had predicted.

The Greek teachings had come to the western world via the Persians. Copernicus was fascinated by their findings, but not convinced by their interpretations. In particular, why was the Sun, a dazzling specter in the sky, so different from the much fainter planets?

Then he had his "eureka, aha" moment, one that would change astronomy forever. From his observations, he deduced that the planets circled the Sun, and not the Earth. If proved, this would remove us from our supposed central place in the Universe, and be an immense blow to religious beliefs. His controversial *De Revolutionibus* was published when he was on his deathbed.

But was Copernicus right? Only better measurements could prove it one way or another. This task was taken up by an eccentric nobleman, Tycho Brahe, who built a palatial castle-cum-observatory on the Danish island of Hven. He boasted a golden nose (it was actually made of brass) and a moose as a pet.

Although he didn't believe in Copernicus' Sun-centered cosmos, Tycho scoured the heavens meticulously. Before the era of telescopes, Tycho built astonishingly complex and accurate measuring instruments at Hven, until he left after the succession of a new King of Denmark, who was eager to rid his country of competitive nobles.

Tycho moved to Prague, where he met with the world's greatest mathematician, Johannes Kepler. The German was the opposite of Tycho, being a shy, sickly man.

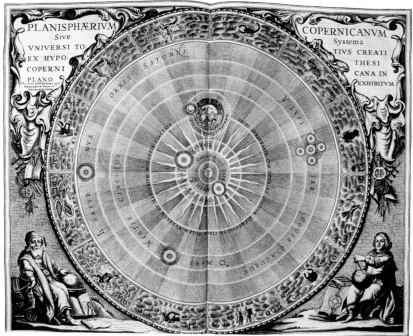

*Copernicus' controversial vision of the cosmos, with the Sun surrounded by circling planets.*

But Kepler was a believer in Copernicus' ideas. Using Tycho's observations, he proved that the planets did indeed orbit the Sun: not in circles, but in egg-shaped orbits called ellipses. When the fast-moving Earth overtakes Mars, for instance, on the inside track, the Red Planet naturally appears to go backwards. There is no need for the complicated Greek wheels within wheels!

Less than a century after Copernicus' revolution in the cosmos, another was poised to begin.

# UNSEEN POWERS: THE TELESCOPE AND GRAVITY

The scene was now set for astronomy to change forever. In 1609, the Italian Galileo Galilei turned his "optick tube"—the newly invented telescope — toward the sky. Galileo quickly announced his findings: that the Earth circled the Sun and that the heavenly bodies were not perfect. For example, the Moon was pocked with craters and the Sun had spots.

This was not good news for the Church authorities, who believed in the supremacy of a central Earth, and the purity of the Sun, Moon and planets.

For teaching Copernicus' ideas, Galileo was placed under house arrest until he died. But his findings inspired a young Englishman, Isaac Newton, who was born in the year of Galileo's death.

*Galileo: pioneer of the telescope*

*Newton: master of gravity*

*William Herschel's mighty 40-foot telescope*

With his formidable mathematical brain, Newton worked out *why* bodies in space move in the way they do. There was an unknown force at work: gravity. At last, astronomers could *calculate* what was going on in the Universe, rather than just predict the future based on what had happened in the past.

Newton was a reclusive figure, and he only published his findings after being pestered by his colleague Edmond Halley. Halley would later use the new theory of gravity to predict the reappearance of a comet that he had seen on his honeymoon in 1682, the one now known as "Halley's Comet."

From Newton's time on, the pace of advancements in astronomy quickened. Some astronomers spent their time trying to prove Newton right or wrong; others employed his new invention, the reflecting telescope (which uses a mirror, instead of a lens, to collect light), to make astonishing breakthroughs.

In the 18th century, the astronomer/ musician William Herschel used a Newtonian telescope to discover a new planet: Uranus. Additionally, by building ever-larger telescopes, he paved the way to exploring the wider Universe, with his investigations into the nature of the Milky Way. He mapped the stars, and came up with the first accurate model of our local galaxy.

# STARS IN THEIR EYES

By the early 19th century, attention was turning to the stars. What were they? What were they made of? What made them shine? And how far away were they?

With precision telescopes, astronomers rapidly answered the last question. By measuring tiny shifts in the apparent positions of stars as the Earth orbits the Sun, they could calculate the stars' distances. The nearest star, Proxima Centauri, turns out to be 4.24 light years away — which means that light from the star, traveling at 300,000 km per *second*, takes over four years to reach us.

The next breakthrough came not from an astronomer, but from a bespoke Bavarian lens-maker, Joseph von Fraunhofer. Testing a new prism, a glass that splits up light into a rainbow, or spectrum, Fraunhofer was horrified to find that the Sun's spectrum was crossed with narrow, vertical lines. Other bright stars showed similar lines. Was his glass at fault?

German chemists Gustav Kirchoff and Robert Bunsen (the inventor of the Bunsen burner) tested out Fraunhofer's results in the lab. They found that the lines corresponded to fingerprints of different elements, including iron, calcium, sodium and carbon.

Meanwhile, in Britain, William Henry Fox Talbot was pioneering the art of photography. Now astronomers

*Cecilia Payne-Gaposchkin*

*A prism splits light into a rainbow of its constituent colors*

had a permanent method of recording the heavens, and the combination of photography and spectroscopy meant that they now knew the make-up of the stars. Or did they?

In the 20th century, it fell to a brilliant British astronomer, Cecilia Payne-Gaposchkin, to discover the truth. By using spectroscopy, she found that stars were made overwhelmingly of hydrogen.

Shortly afterwards, Arthur Eddington, the leading British astronomer of his time, came up with the answer. Stars were giant nuclear reactors, fusing hydrogen into helium in their super-hot cores. The fusion reactions created the energy that made the stars shine.

But where did the other elements come from? It was the British/American team of Margaret and Geoffrey Burbidge, Willy Fowler and Fred Hoyle (fondly known as $B^2FH$) who calculated that stars more massive than the Sun have the gravitational power to create heavier and heavier elements in their cores — until they try to fuse iron, when the process breaks down spectacularly.

That is when the massive star explodes as a supernova, spewing its debris over the cosmos, and seeding the new generation of stars and planets. Its dying core may turn into a neutron star, or even a black hole.

# THE UNIVERSE UNVEILED

Now that astronomers had worked out the nature, distances and composition of the stars, it was time to tackle a mega-question: the structure of the distant Universe.

Was our Galaxy, which William Herschel had mapped, all that existed or was it just one of billions of galaxies? The latter proved to be the case. You can see the evidence for yourself, with the unaided eye. In the southern hemisphere are the two stunning Magellanic Clouds: satellites of our Milky Way. In the northern hemisphere, there's the great Andromeda Galaxy.

Later, in one of the greatest discoveries of the last century, American astronomer Edwin Hubble found that the entire Universe is expanding. As a result of the Big Bang, a colossal cosmic explosion that took place 13.8 billion years ago, the galaxies are all flying apart from each other.

The proof of this cataclysm came in 1965 when American physicists Arno Penzias and Robert Wilson detected radiation from the inferno all over the sky — the "afterglow" of the Big Bang.

The Universe is not just expanding; in the 1990s astronomers discovered that the rate of expansion is accelerating. The reason for this is a force known as "dark energy." What is it? No one knows. But the future of our cosmos will become colder, lonelier and darker (see pages 274–5).

Very recently, a new revolution in astronomy has taken place — one as great as the upheaval in the era of Copernicus and Galileo. We can now send robotic explorers to the other planets. Moreover, new technologies mean that astronomers are no longer limited to simply *looking* at the sky. They can now tune into the cosmos at a whole range of wavelengths, from hugely energetic gamma rays to low-frequency radio waves.

This recent wealth of data on the cosmos has told us that we live in a violent universe. The safe, predictable stars

*Depths of the Universe — a swarm of distant galaxies.*
*The star is in the foreground.*

and planets of our ancestors have been replaced by wild worlds. Black holes, quasars, wayward planets and exploding stars are all out there on view.

However, astronomers are also accruing evidence that, despite all this disruption, there could be life somewhere else out there in the cosmos.

# CHAPTER 2
# OBSERVING THE SKY

# INTRODUCTION

You're an astronomer already, even if you don't know it. Every time you see the Moon, spot the brilliant "Evening Star" or catch a glimpse of a shooting star, you are making an astronomical observation. And there is an awful lot of astronomy that you can do with the naked eye (see pages 34–5).

The best way to start out in astronomy, in fact, is not with a computer-controlled telescope, but just using your eyes to find your way around the sky. Get your bearings with our seasonal charts found on pages 28–35.

Locate the brightest stars. Then trace out the star patterns, each of which is unique and has its own fascinating history (there is a detailed constellation-by-constellation guide on pages 276–369). Follow the changing phases of the Moon and the path it traces through the stars every month, through the Zodiac. If you see an intruder among the star patterns, it is likely to be a planet. Except for Uranus and Neptune, the planets are so bright you can easily see them with the naked eye; Venus and Jupiter are more brilliant than any of the stars in the night sky. One easy-to-spot difference is that while stars twinkle, the planets shine with a steady glow. Check our listings on pages 384–387 for the current date, to check which planets you're seeing.

Then, explore the sky in more detail through a pair of binoculars. Using their wide field of view and modest magnification, you can learn your way around the sky, hopping from star to star to track down glowing nebulae, scintillating clusters of stars and distant galaxies.

The ultimate accessory is a telescope. The word "astronomer" probably conjures up the image of a white-bearded man peering through a long telescope. Although today's astronomer is as likely to be female as male, a telescope is still essential if you want to take your hobby to the limit. Your telescope will open up all the glories of the night sky, with the opportunity to image the depths of the Universe for yourself.

*For more information on equipment manufacturers and useful websites, check out the links on page 391.*

*Anyone can be an astronomer. A warmly-dressed observer uses a small telescope to scan the sky.*

# GETTING STARTED

Great, you are going out for a star-gazing session. It is best not to be stumbling about in the dark, though, so first check out your observing site during the day, even if it is your own yard.

You will need a spot with a good view of the sky, especially in the direction where the Sun is highest at midday, as that is generally where you'll find the Moon and planets at night. Remember this direction so you can orientate your star chart later. (This is south if you're in the northern hemisphere; north if you are below the equator.) Avoid areas of grass, as they are

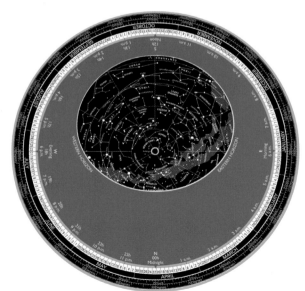

*A planisphere shows the positions of the stars and constellations for any hour in the year. This particular one is designed to be used at latitude 51.5° North (for example, in northern Europe, northern USA and Canada). If you live further south than this, you will need to use a planisphere that is suitable for your own latitude.*

damp and cold at night, and anywhere that is in the direct glare of streetlights.

## Observing kit

Dress warmly. Even in summer, you can get surprisingly cold when standing still. Wear two pairs of socks, layers of clothes and a warm hat. Use a red-light flashlight to read your star charts and make notes (bright white light spoils your night vision — see Dark Adaptation panel).

Use the charts on the following pages to find your way around the sky. Turn to page 28 if you are in the northern hemisphere, or page 32 if you are south of the Equator.

Alternatively, you can buy a planisphere (make sure that it is for your latitude) and set it for the current date and time. Alternatively, download an app that shows the sky view in whichever direction you point your smartphone or tablet.

Enjoy the naked-eye view before turning to binoculars or a telescope. Your eyes are actually the best astronomical instruments for observing large-scale sky sights, like the Aurora Borealis and Australis (the Northern and Southern Lights), showers of meteors or the whole sweep of a constellation like Orion.

## DARK ADAPTATION

When you first go out at night, you will see only the brighter stars. Do not be disappointed. Give your eyes half an hour to adapt, and they will become sensitive enough to spot hundreds or thousands of fainter stars. There are two reasons. First, the eye's pupil (the hole in the center of the colored iris) expands in darkness to admit more light. Second, and more importantly, in dark conditions, the retina at the back of the eye builds up more visual purple (rhodopsin), a chemical that responds to light, and this allows you to see at lower light levels.

# FINDER CHARTS

Our outlook on the night sky depends on the time of year; as the Earth travels around the Sun, we look out to different regions of space. It is like the changing view from a rotating carousel at the fair; one moment we see a roller coaster, next the bumper cars.

We see the great hunter Orion during the winter months in the northern hemisphere and during the summer months in the southern hemisphere. Six months later, when we are on the far side of the Sun, Orion is nowhere to be seen and our sky is graced with the lovely constellations Lyra and Cygnus in the northern hemisphere or the brilliant celestial scorpion, Scorpius in the southern hemisphere. (The southern hemisphere's iconic Southern Cross, Crux, puts on its best show in autumn.)

## SEASONAL CHARTS, NORTHERN HEMISPHERE

These three pairs of charts (pages 29–31)

show the evening sky, looking south and north, as you see it in January, May and September. The yellow "Ecliptic" line is the path followed by the Sun, Moon and planets.

We have picked out the brightest stars and the major constellation patterns. There is enough overlap in the maps for you to be able to work out the view in the intervening months, too. For instance, in the south-facing charts, Leo is at the left (east) in January and to the right (west) in May, so in February and March it is roughly in the middle, due south.

The bottom edge of the south-facing charts depicts the horizon, as seen if you are at a latitude of 30 degrees north. Because the Earth is curved, people nearer to the North Pole can see less of the southern region of the sky: the three curved pink lines show your horizon if you are 40, 50 or 60 degrees north.

Conversely, people at higher latitudes can see more of the northern sky. The bottom edge of the north-facing charts

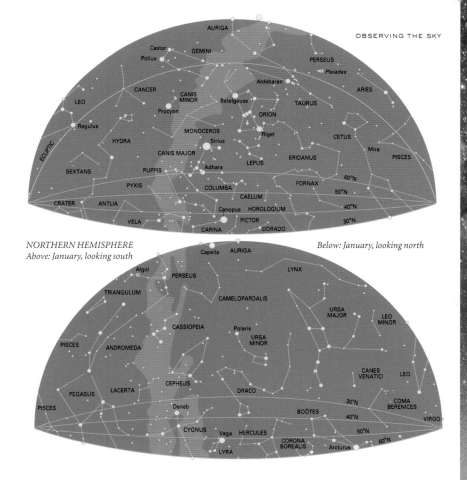

*NORTHERN HEMISPHERE*
*Above: January, looking south*

*Below: January, looking north*

shows the horizon for 60 degrees north; successive pink lines indicate the horizon if you are 50, 40 or 30 degrees north.

## Morning skies

If you are out very early in the evening or up before sunrise, you will notice that

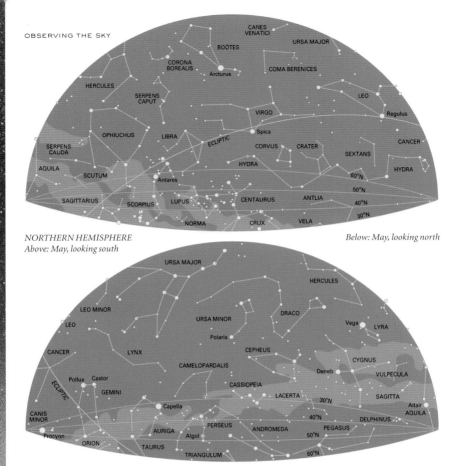

*NORTHERN HEMISPHERE*
*Above: May, looking south*

*Below: May, looking north*

the positions of the constellations do not match the charts on these pages. That is because the Earth is spinning around on its axis, every 24 hours, which seems to make the sky move around the opposite way.

The stars, planets and Moon appear to rise in the east and set in the west. To extend our carousel analogy, it is as if the horses are turning on their own poles, as well as cantering around its rim.

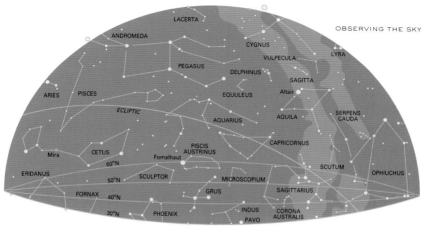

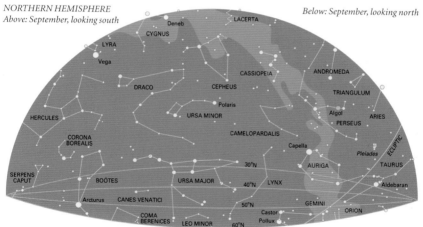

*NORTHERN HEMISPHERE*
*Above: September, looking south*

*Below: September, looking north*

Roughly speaking, the charts for January evenings apply also to October mornings; those for May evenings to February mornings, and those for September evenings to June mornings.

Do not be daunted by the rather complex movements in the night sky. Once you are familiar with some of the brighter constellations and stars, such as Orion, Leo, Arcturus, Antares, the Ursa

Major and Cassiopeia (for those of you in the northern hemisphere) and the Southern Cross, Orion, Leo, Canopus, Scorpius and Arcturus (for those of you in the southern hemisphere), you will be able to find your way around as easily as using the major landmarks to navigate around your local neighborhood on Earth.

## SEASONAL CHARTS, SOUTHERN HEMISPHERE

These three pairs of charts (see pages 33–35) show the evening sky, looking north and south, as you see it in January, May and September. The yellow "Ecliptic" line is the path followed by the Sun, Moon and planets.

We have picked out the brightest stars and the major constellation patterns. There is enough overlap in the maps for you to be able to work out the view in the intervening months, too. For instance, in the north-facing charts, Leo is at the right (east) in January and to the left (west) in May, so in February and March it is roughly in the middle, due north.

The bottom edge of the north-facing charts depicts the horizon, as seen if you are at a latitude of 5 degrees south. Because the Earth is curved, people nearer to the South Pole can see less of the northern region of the sky: the three curved pink lines show your horizon if you are 15, 25 or 35 degrees south.

Conversely, people nearer the South Pole can see more of the southern sky. The bottom edge of the south-facing charts shows the horizon for 35 degrees south, while successive pink lines indicate the horizon if you are 25, 15 or 5 degrees south.

## GET SET AND GO

Now you are ready to observe. We suggest that you skip the rest of this chapter on first reading; after you get to know the sky, come back and learn more about the instruments that will take you further.

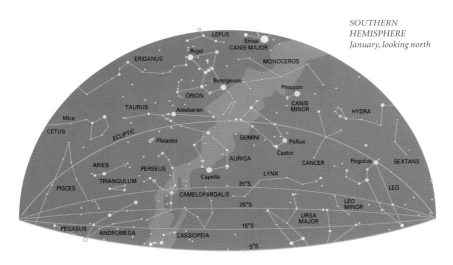

*SOUTHERN HEMISPHERE*
*January, looking north*

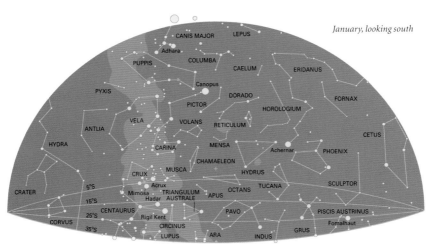

*January, looking south*

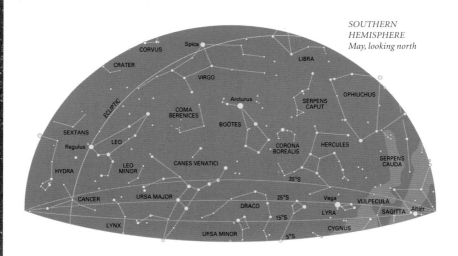

*SOUTHERN
HEMISPHERE
May, looking north*

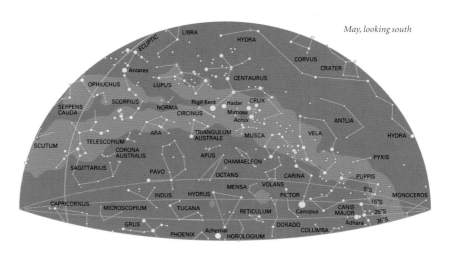

*May, looking south*

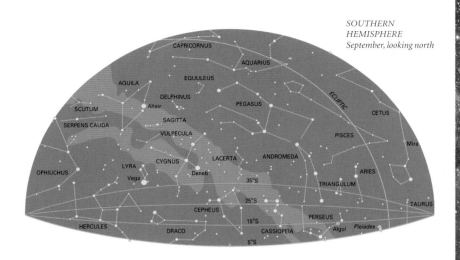

*SOUTHERN HEMISPHERE*
*September, looking north*

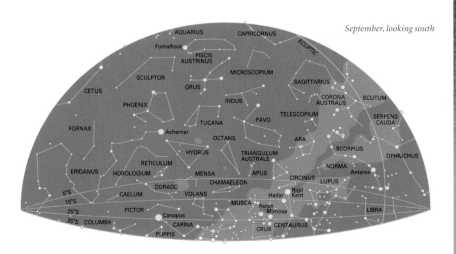

*September, looking south*

35

# BINOCULARS

Before you buy a telescope, invest in a good pair of binoculars. This is not just a cut-rate alternative; most serious astronomers use binoculars as well as their main telescopes. These handy little instruments are ideal for many purposes where you do not need huge magnification; they also show you a wide swathe of sky.

Binoculars will give you stunning views of star clusters, and the nebulae and star clouds scattered along the Milky Way.

*Binoculars are great for stargazing — and for many outdoor pursuits.*

You can easily spot Jupiter's moons, as well as the most prominent galaxies. If there is a bright comet around, binoculars usually provide the best views of the tail.

## EASY TO USE

Moreover, binoculars are lightweight and easy to hold; unlike a telescope, you do not need a massive mounting to support them. Binoculars are also much easier to use because they show the view the right way up, while astronomical telescopes show everything upside-down.

It will help if you can support the binoculars, otherwise they will magnify every tiny shake of your hands. Generally, it is good enough to prop them on a fence or just rest your elbows on a table. Alternatively, you could buy a pair of image-stabilized binoculars. They are expensive, but it is amazing how much more you can see when the stars are not wobbling around.

## BUYING BINOCULARS

If you can, buy binoculars from a shop rather than online, so you can test them to make sure that you are happy with their weight, magnification and performance.

Binoculars come with a designation, such as 7x50 or 12x70. The first figure is the magnification; the second is the size of the main lens (in millimeters). With binoculars, it is not true that "bigger is better." If the magnification is too high, this will accentuate your shaking as you hold the binoculars. Added to that, binoculars with bigger lenses are much heavier; either your arms will tire, or you will need a mounting.

We recommend 7x50, or perhaps 10x50, as the ideal binoculars for astronomy.

# TELESCOPES

When American astronomers planned a large observatory on Kitt Peak in Arizona, they had to persuade the local Tohono O'odham people to allow them access to their sacred mountain. The scientists invited the tribal elders to view the Moon through a telescope. The Native Americans were so impressed that they immediately granted permission to "the Men with the Long Eyes."

Meaning "far-seeing" in Greek, a telescope really does bring the heavens down to Earth. You can swoop over the craters of the Moon, see the ringed planet Saturn suspended in space like an exquisite model and view distant galaxies.

Over the next few pages, we cover the three main kinds of telescope, with their pros and cons, along with the equipment you will need to use with them, such as eyepieces, finders and mountings. If you acquire a compact telescope for casual viewing, you can usually just carry it outdoors (but store it in a cool

## ABOVE THE ATMOSPHERE

However good your telescope, you always have to look upwards through Earth's churning atmosphere, which makes everything shimmer, rather like looking up from the bottom of a swimming pool. That is why professional astronomers now site their telescopes on high mountain peaks above the worst of the air's turbulence. Space telescopes, such as the Hubble Space Telescope, and its successor the James Webb Space Telescope, observe from space, where they are completely clear of the atmosphere.

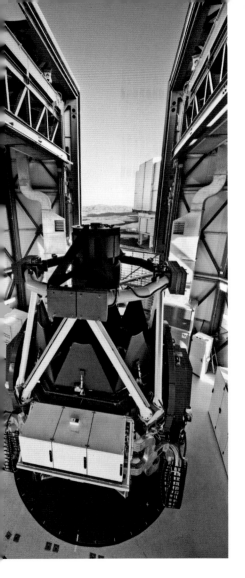

place to avoid it dewing over). If you are serious enough to set up a telescope on a permanent mounting, you will need to protect it with either a roll-off shed or an observatory building with a slit in the roof.

### BIGGER IS BETTER

When acquiring a telescope, go for the biggest you can afford. First, it will allow you higher magnification (see eyepieces and magnification on page 43). A high magnification on a small telescope will just give you a larger blurred image, what astronomers call "empty magnification." The maximum magnification you should use is twice the telescope's diameter in millimeters, so there is no point in "pushing" a 75 mm telescope beyond a magnification of 150 times.

Second, the bigger a telescope, the greater its "light grasp." A small telescope will reveal Uranus and Neptune, but if you want a good view of much dimmer objects, such as remote galaxies, a large telescope is essential.

*The VLT Survey Telescope scans the sky from a 2,635 m mountain in Chile's Atacama Desert.*

# REFRACTORS

The traditional telescope, the kind used by sailors and early astronomers, has a large lens at the front end of its tube to gather light and bring it to a focus. A smaller lens (the eyepiece) at the other end magnifies the image. Because the main lens bends, or "refracts," the light, this type of telescope is called a refractor.

This was the earliest type of telescope, and was invented by optician Hans Lipperhey in Holland in 1608. Galileo was one of the first to observe the sky with the new type of instrument.

The front lens of the telescope (known as the "objective") not only focuses the light but also spreads it out into a multi-hued rainbow, in an effect called chromatic aberration. If you look through a very cheap refractor, you will see a star surrounded by a blurry colored fringe.

To make useful observations, astronomers had to develop a refractor in which the objective is made from two lenses, composed of different kinds of glass, stuck together to reduce the color fringing. All large professional refractors are made using this *achromatic* design.

During the 19th century, astronomers

*Two of Galileo's tiny refractors*

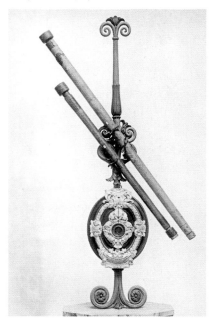

*The Yerkes Telescope in Williams Bay, Wisconsin, is the largest refractor in the world. Its 40-inch (1 m) lens is the largest practicable; any bigger, and it would sag under its own weight.*

competed to build bigger and bigger refractors, culminating in the giant telescope at the Yerkes Observatory, near Chicago. However, its lens, at 40 inches (1.02 m) across, is about as large as you can make a refractor; a bigger lens would sag under its own weight, and would not focus properly. That is why, apart from special telescopes for observing the Sun, astronomers have not constructed a large refractor for over a century.

### CHOOSING A REFRACTOR

A refractor is a very good telescope for backyard stargazing. Although more expensive than a reflector of the same

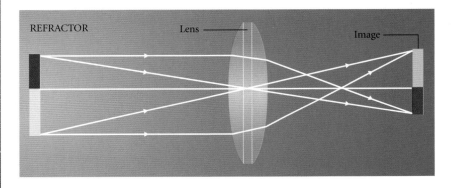

REFRACTOR      Lens

Image

diameter, a decent refractor is simple and fairly rugged, and gives sharp views with high contrast, so it is ideal for observing the Moon and planets.

Do not buy the cheapest you can find in a catalog or store. These will have only a single objective lens. To reduce the chromatic aberration that results from having only one lense, manufacturers of cheaper refractors install a plastic ring that is designed to cut down the resulting chromatic aberration, but it also ends up blocking out most of the light coming into the telescope!

You will need to choose an achromatic refractor (also described as having a "doublet lens") or, if your budget will allow it, a top-of-the-line *apochromatic*

reflector, in which the objective consists of three lenses and eliminates chromatic aberration altogether.

*Using a refractor to observe — note the sturdy mount*

# EYEPIECES AND MAGNIFICATION

To see the view through any telescope, you need a set of different eyepieces that give a range of magnifications. When you observe a planet like Saturn, for example, you will want to start with the lowest-power eyepiece for a general view, then switch to eyepieces that give you higher and higher magnifications.

As a general rule, the biggest eyepieces in your set will give you the lowest magnification, while the smallest have the highest power.

You can work out the magnification as follows: in the specifications for the telescope, look for the focal length of the objective lens. Each eyepiece has its focal length marked on it. Divide this into the focal length of the telescope lens, and the result is the magnification.

For instance, if your telescope has a focal length of 500 mm, and you select a 10 mm eyepiece, you will see Saturn magnified by 50 times (500/10).

## THE WORLD'S LARGEST REFRACTORS

| NAME, LOCATION | DATE COMPLETED | DIAMETER OF LENS (M) |
|---|---|---|
| Yerkes Refractor, Wisconsin | 1897 | 1.02 |
| Swedish Solar Telescope, Canary Islands, Spain | 2002 | 0.98 |
| Lick Telescope, California | 1888 | 0.91 |
| Grande Lunette, Meudon, France | 1891 | 0.83 |
| Grosse Refraktor, Potsdam, Germany | 1899 | 0.80 |
| Grande Lunette, Nice, France | 1886 | 0.77 |

# REFLECTORS

A reflecting telescope, or reflector, uses a mirror, instead of a lens, to reflect light to a focus. Unlike refractors, reflectors do not suffer from chromatic aberration. Isaac Newton built the first reflector in 1668. Newton also inserted a second small mirror at an angle to reflect the focused light onto an eyepiece (see page 43) on the side of the telescope.

In the biggest telescopes of the 20th century, an astronomer would sit at the focus for cold, uncomfortable hours of observation, inside the telescope itself! Now that professionals make observations

*A reflector is usually the most popular choice for an amateur astronomer. You can even make one yourself, grinding your own mirror. The weight of the mirror is supported from the back, so there's no size limit for a backyard telescope.*

with electronic devices, you will find the astronomer at a computer in the warm control room.

A reflector has the great advantage that you can support even the largest mirror from behind to prevent it from sagging under its own weight. That is why all the biggest telescopes today are reflectors. Many have mirrors supported by computer-controlled actuators, which are small devices that push on the mirrors to keep them exactly the right shape as the telescope tilts.

The Keck 1 Telescope on Hawaii pioneered the way to such colossal sizes; it collects light with a set of 36 hexagonal mirrors that fit together precisely, like bathroom tiles, to create a mirror surface larger than you could create from a single piece of glass.

## FINDERS

So you have a telescope and you want to point it at Jupiter and find that it is surprisingly tricky. The view through the eyepiece is unexpectedly restricted, so aiming the 'scope is a bit hit-and-miss.

This is when you need a finder. Traditionally, a finder is a small telescope pointing in the same direction as your main instrument, but with a low magnification and a wide field of view. If you find Jupiter in the little 'scope and center it up on the crosswires, you will immediately see it in your main eyepiece, too. Many telescopes now come instead with a red dot finder, which superimposes a red dot on the sky where the telescope is pointing.

With either type, check that it lines up properly with your main telescope by looking at a distant object during the day (when it is much easier to make adjustments).

## NEWTONIAN REFLECTOR

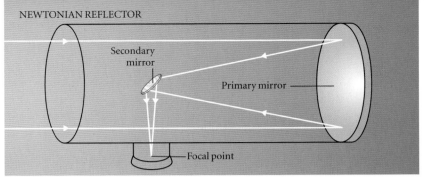

Secondary mirror

Primary mirror

Focal point

### CHOOSING A REFLECTOR

Size for size, a Newtonian reflector is the cheapest backyard telescope you can buy, and shows you the faintest object. A reflector will give fantastic views of glowing nebulae and distant galaxies.

A Dobsonian telescope is the ultimate "light-bucket," a huge Newtonian telescope resting on Teflon pads in a simple swivelling framework; you just push it to view the very faintest sky sights.

However, reflectors need more maintenance than refractors. They go out of adjustment more easily, and the mirror's reflecting surface needs recoating with aluminum every few years. In addition, the small mirror in the tube obstructs some of the light so the view is not entirely crisp, which is a drawback if you want to observe the Moon and planets in detail.

*(Above) Reflectors gather light with a large mirror, then deflect it to a Secondary mirror that feeds it to the focus.*

*(Below) One of Newton's first reflectors, constructed in 1668*

*A Keck Telescope on Hawaii. Its 10 m mirror is made of hexagonal segments, creating an enormous glass surface.*

## THE WORLD'S LARGEST REFLECTORS

| NAME, LOCATION | DATE COMPLETED | DIAMETER OF MIRROR (M) |
| --- | --- | --- |
| Gran Telescopio Canarias, Canary Islands, Spain | 2009 | 10.4 (segmented) |
| Keck 1 & Keck 2, Hawaii | 1993, 1996 | 10 (segmented) |
| South African Large Telescope, South Africa | 2005 | 9.2 (segmented) |
| Large Binocular Telescope, Arizona | 2004 | 8.4 (2 mirrors) |
| Very Large Telescope, Paranal, Chile | 1998-2001 | 8.3 (4 telescopes) |
| Subaru, Hawaii | 1999 | 8.3 |

# CATADIOPTRIC TELESCOPES

In 1930, Estonian optician Bernhard Schmidt devised a new kind of telescope. This powerful new tool for probing the depths of space has revolutionized backyard stargazing. Remarkably, Schmidt had lost a hand at age 15, experimenting with gunpowder.

Look through a traditional telescope, refractor or reflector, and you suffer from tunnel vision; it gives you only a narrow view of the sky. However, what Schmidt did was to combine these two designs to create the first wide-view telescope.

The Schmidt telescope has a large

*Cadatioptric telescopes use both lenses and mirrors to gather light. They have a wide field of view.*

mirror at the bottom that focuses light, but on its own would give fuzzy images at the edge of the field of view. Schmidt's stroke of genius was to put a specially shaped thin lens at the top of the telescope tube to sharpen the images.

## MOUNTINGS

Whatever kind of telescope you have, you need a mounting to support its weight and to allow you to aim at whatever catches your attention in the heavens. The complication is that objects in the sky seem to move continuously as the Earth rotates, so you need to track whatever you are viewing.

In the simplest mounting, known as an alt-az, your telescope swings between the prongs of a rotating fork, a design that allows to you swivel the telescope in altitude (up-and-down) and azimuth (around). If you are using a low magnification, you can track an object simply by pushing the telescope.

Most catadioptric telescopes come with an alt-az mounting that is equipped with computer control. Once you have set up the telescope using two bright stars, motors within the mounting will follow your chosen object automatically. These are known as GO TO mountings and have databases of thousands of astronomical objects. You can simply key in your chosen target and the telescope will automatically slew to find it.

Unfortunately, that is no use for astrophotography. The view gradually rotates as the telescope tracks across the sky, thus blurring your image.

The alternative, and ideal for imaging, is an equatorial mounting. This is tilted over at an angle to match your latitude on Earth. Although it is tricky to set up, an equatorially-mounted telescope needs only one motor driving at a constant speed to keep your view constant.

The focused light is recorded on a photographic plate within the telescope. In the late 20th century, large Schmidt telescopes mapped the far Universe in unprecedented detail, discovering, for instance, vast clusters of galaxies.

For backyard stargazers, the disadvantage is that you cannot look at the sky directly through a Schmidt telescope. However, in a related design, the Schmidt-Cassegrain telescope (SCT), uses a small mirror within the tube to focus the light onto an eyepiece at the back. The Maksutov is a similar design, with a dish-shaped lens at the front. These are called catadioptric telescopes, combining the power of a mirror (catoptric) with a lens (dioptric).

## CHOOSING A "CAT"

A catadioptric telescope is a lot shorter and lighter than a refractor or reflector of the same diameter, so it is easier to use and carry around. A major drawback, however, is that it is also more expensive. The large central mirror means that the images are not as sharp as you would see in a refractor, but it is ideal for viewing deep-sky objects like nebulae and galaxies and is now the most popular design for serious amateur astronomers.

*The optics of a catadioptric telescope — lens (left) and mirror (right) — make for a very compact instrument.*

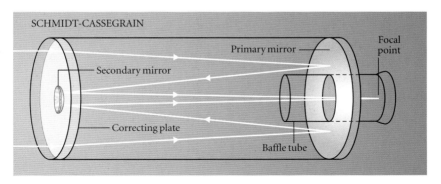

SCHMIDT-CASSEGRAIN

Primary mirror

Focal point

Secondary mirror

Correcting plate

Baffle tube

*The UK Schmidt Telescope at the Australian Astronomical Observatory, New South Wales. Professional cadatioptric telescopes are named in honor of Bernhard Schmidt. Their wide field of view makes them excellent for undertaking sky surveys.*

## THE WORLD'S LARGEST SCHMIDT TELESCOPES

| NAME, LOCATION | DATE COMPLETED | DIAMETER OF LENS (M) |
|---|---|---|
| Alfred-Jensch-Teleskop, Tautenburg, Germany | 1960 | 1.34 |
| Samuel Oschin Telescope, California | 1948 | 1.22 |
| UK Schmidt Telescope, Siding Spring, Australia | 1973 | 1.2 |
| Calar Alto Telescope, Spain (originally Hamburg, Germany) | 1980 (1955) | 0.8 |

# IMAGING THE SKY

We have all seen fantastic pictures of the sky taken through big telescopes, but imaging the sky is not easy. Point your standard camera at the night sky and click, and the result will be a blank image; the stars and planets are too faint to show up. You might catch the Moon, although it will be a disappointingly small speck of light.

(Even worse, your flash will probably go off automatically and light up the foreground.)

### STEADY AS YOU GO

Check that your camera can take long exposures (often marked as a B-setting). The camera must be rock-steady to prevent blurring. Ideally, use a cable

*Under clear desert skies, observers line up telescopes and cameras for an evening shoot.*

*Exquisitely-detailed image of giant planet Jupiter, captured by an amateur astrophotographer. Its moon Io is top left.*

release for the shutter, or a self-timer, so your finger does not shake the camera.

An ordinary camera tripod (or even holding the camera firmly on a wall) allows you to capture reasonable views of the Moon and planets. Remember, the more you zoom, the steadier the mounting has to be, and the fainter the final image will turn out.

On a wide-angle setting, you can image large-scale sky sights, like the aurorae and noctilucent clouds. Leave the shutter open for several hours, and you can acquire amazing shots of the "star trails" traced by stars wheeling across the sky as the Earth rotates.

But the Earth's rotation also means that your long exposures of all celestial bodies will be stretched out. To get a sharp image, you need a mounting that tracks these objects. If you have such a telescope mounting, you can attach your camera to that; in fact, you can buy a motor-driven mounting just for the camera. Using a wide-

angle setting, you can get excellent views of the constellations and the Milky Way.

## THROUGH THE TELESCOPE

If you have a telescope, you can simply hold your camera (or smartphone) up to the eyepiece and snap away. With luck, you can obtain fairly good images of the Moon and planets.

Even better, if you have a digital SLR (DSLR) camera, buy an adaptor (they're fairly cheap), take out the eyepiece and attach the camera directly to the telescope. Because Earth's shifting atmosphere blurs the image, you may need to take quite a few pictures before you obtain one that is totally sharp.

This is why many serious amateurs now use webcams on their telescopes instead of cameras. Several are available that are specially adapted for astronomy. They are not expensive (in comparison with your telescope). Webcams take many pictures per second and you don't have to check through them all yourself to

*Time-lapse image of astronomers with flashlights at an astrophotography session, Isle of Wight, UK.*

find the best; there are several software programs that will automatically identify the sharpest images and even add them together to create a stunning picture of the Moon or planets.

## DEEP SKY IMAGING

Fainter objects, like nebulae and galaxies, need longer exposures, from a minute up to several hours on your telescope. It is no surprise that astronomers call them "deep sky objects."

Your telescope must track the object accurately across the sky, so you will need a good motorized mounting (see page 49). Even so, the images will probably drift a bit during a long exposure, so you have either to spend hours at the eyepiece, correcting any unwanted motion, or to invest in an autoguider that will correct the telescope drive automatically. You can use a DSLR camera, but if you are going to this much effort, it is better to use a light-sensitive electronic chip, known as a CCD (charged-coupled device). You can buy these from a telescope shop and they give stunning results.

# REMOTE OBSERVING

Imagine if you could view the planets, nebulae and galaxies with a telescope much bigger than you could have in your backyard. These days, that is no pipe dream, but is open to everyone, thanks to organizations that have set up instruments around the world for you to operate from the comfort of your own computer.

The wide range of locations means that you can observe your favorite objects if it is cloudy outside where you are, or even in daytime. In addition, you can image objects in the other hemisphere, which cannot be seen at all from your home.

Usually, you need to book time on these telescopes, and there is usually a fee to pay. Once you have started observing, the excitement builds up until the distant telescope lets you know that your exclusive image is ready for you to access online.

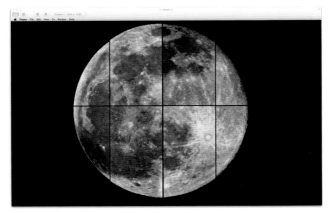

*Observing at a distance: computer screen connected to one of the giant telescopes at the Kitt Peak National Observatory in Arizona.*

## LIGHT POLLUTION

We once received a letter from a lady who asked, "Before the war, we could see so many stars. But they're not there any more. Have they faded?"

"No" is the definite answer. The culprit is light pollution: stray illumination from streetlights, security lights and sports stadiums is making the sky above our cities and towns glow so brightly that the fainter stars are lost from view.

Light pollution also disturbs nocturnal animals and birds. It wastes electricity, costing us all money and adding to global climate change. The International Dark-Sky Association is leading the campaign for better lighting at night, and is endorsing Dark-Sky Parks around the world where we can all enjoy the true beauty of the night sky.

### ASTRONOMY HOLIDAYS

Another way to use a big telescope, hands-on this time, is to book a holiday at a residential astronomy center. Here, the hosts not only provide accommodation and meals, but will also show you the sky sights through a large telescope, demonstrate how to take pictures with their equipment and often allow you to use your own camera on their telescopes.

Additionally, astronomy centers are usually located in remote areas where the sky is really dark, so there are plenty of interesting sites to visit during the day for astronomical and non-stargazing members of the family alike.

# RADIO ASTRONOMY

At the height of the Second World War, the British military was alarmed by powerful radio signals that were jamming their radar sets. A young scientist, Stanley Hey, quickly realized that the perpetrator was in fact an immense magnetic storm, on the Sun.

The discovery showed that visible light is not the only kind of radiation coming from space. Following Hey's discovery, astronomers built huge radio telescopes, like that at Jodrell Bank, England; Parkes, Australia and the Very Large Array, New Mexico, to garner radio waves from the sky. They reveal a cosmos full of violence, one that is invisible to ordinary telescopes.

Radio astronomers have discovered powerful radio waves from exploded

*The 76 meter diameter Lovell Radio Telescope at Jodrell Bank, Cheshire, UK. This huge dish began work in 1957.*

stars, pulsars and distant erupting galaxies (including the enigmatic quasars). Radio waves generated at the birth of the Universe also criss-cross space; some of the interference you see on a television set that is not properly tuned is in fact radiation from the Big Bang.

*Biggest radio telescope in the world: the 305-meter Arecibo dish in Puerto Rico*

## BUILD YOUR OWN RADIO TELESCOPE

If you are reasonably proficient in electronics, you can make your own radio telescope. Search on the Internet for a design that suits you.

You will need an antenna to collect the signals from space, either a satellite television dish or a carefully arranged set of wires. Link it to a detector, which tunes into the radio waves, and an amplifier to boost the signal, and you have your own cosmic radio receiver.

Scan your radio telescope across the sky, and hunt for radio sources. If you are listening in, you will hear a burst of noise, while the dial on your receiver shows the strength of the signal. You can rig up a chart-recorder for a paper trace or log the results on your computer.

You will be able to pick up radio waves from the Milky Way, as well as from storms on the Sun and around Jupiter.

# BEYOND THE VISIBLE

Light and radio waves are not the only messengers that head our way from exotic objects across the cosmos; there is a whole band of other kinds of radiation that together make up the electromagnetic spectrum.

They all travel at the speed of light, like waves spreading across space. The form of radiation with the longest wavelength (radio waves) vibrates with the lowest frequency, while the short-wavelength gamma rays vibrate most quickly.

Think of it as a piano keyboard, but, instead of sound, it is filled with various kinds of radiation. Ordinary visible light occupies a small range in the center, just middle C and the notes to either side.

Radio waves are the very low frequencies, the deep notes at the far left of the keyboard. In between, we find infrared ("heat radiation"), which is very familiar to us from thermal images of our homes showing where heat is leaking out. For astronomers, infrared telescopes are key in picking out where new stars are being born deep within dark interstellar clouds.

Moving up the scale of radiation from visible light to higher frequencies, we have ultraviolet rays (the ones responsible for suntans) and then X-rays. The highest "notes" of all are gamma rays.

The Earth's atmosphere blocks most of the high-frequency forms of radiation from space, so astronomers have launched satellites to detect them. They have discovered such phenomena as colliding magnetic loops on the Sun, vast clouds of gas at a temperature of millions of degrees and gas that is on the brink of falling into a black hole.

*The Japanese satellite Hinode looks at the Sun and the powerful X-rays it emits. This view of the lower corona — our star's atmosphere — shows bright magnetic loops, which seethe with violent activity, spewing charged particles into space.*

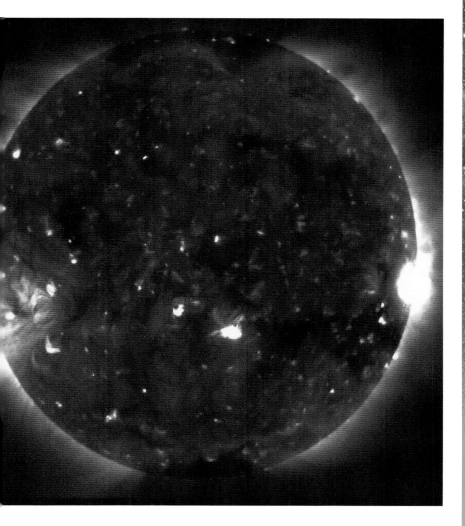

# CITIZEN SCIENCE

Perhaps you are more an "armchair astronomer" than someone who wants to go out and observe on a cold, dark night?

Never fear. In the past few years, professional astronomers have reached out to people like you, to help them solve the puzzles of the Universe.

Now there are dozens of "citizen science" projects, in which you can log on and use your computer power, and your brain, to help analyse anything from the migration of amphibians to the shape of protein molecules in living cells. However, the most popular projects are in astronomy.

Some of these citizen science projects simply run in the background, when your computer is idle. Without any effort from you, except for downloading the program, the software can work out the shape of asteroids, plot the stars in the Milky Way and even search for radio signals from alien civilizations.

## BE A PROFESSIONAL

It is more fun to join in yourself. Typically, a citizen science project will distribute spacecraft and telescope images that are saturated with far more data than the professional scientists can handle.

After some guidance, you are free to inspect thousands of astronomical images, most never seen before by any human being, and classify them by their size, shape and location. If you are a fan of the Solar System, you can check out craters on the Moon or clouds on Mars. You can search for planets orbiting other stars or measure the distribution of mysterious "dark matter" in the depths of space.

These projects are not just limited to visible-light images. You can download pictures of the Sun taken at ultraviolet wavelengths to study magnetic explosions on our nearest star.

You can even take part in a survey to discover where baby stars are being born, right now, in our Galaxy. At the end of

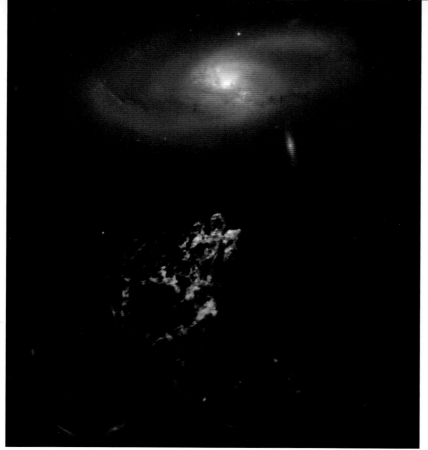

*Discovered by Dutch schoolteacher Hanny van Arkel as part of the Galaxy Zoo project, "Hanny's Object" (green) is a huge stream of gas torn from the galaxy above, and now spawning stars.*

the day, the Universe is yours to explore: whether you are just eyeballing the beauty of the night sky, using your own eqipment to probe deeper into space or connecting with the world's most powerful telescopes to explore the ultimate secrets of the cosmos.

# CHAPTER 3

# THE MOON

# INTRODUCTION

Well before we recognize anything else in the night sky, we all know the Moon. The familiar features of the Man in the Moon have looked down on us since childhood. In western tradition, the Moon is the beautiful and chaste Diana, goddess of hunting. However, in many parts of the world, the Moon is a male deity; in Hindu mythology, it is the god Soma, who rides through the sky in a chariot pulled by white horses.

When astronomers first turned telescopes on the Moon in the early 17th century, they saw it was a world rather like the Earth, but very rough and mountainous. The early English astronomer Sir Thomas Lower said that it appeared "like a tart that my cook made me last week; here a vein of bright stuff, and there of dark and so confusedly all over."

In fact, the Moon is a dry ball of rock, with a diameter of more than a quarter of the Earth. That makes it unique in the Solar System. Many other planets have moons, but they are all tiny compared to their parent planet. The Moon is so large in comparison to Earth, that astronomers often call the Earth-Moon system a "double planet." The Moon was probably formed in a giant cosmic collision when another planet hit the early Earth.

*Buzz Aldrin on the Moon*

## THE DOWDY MOON

The Moon looks amazingly bright in our night sky, but it is actually a very dull object. When astronauts landed there, they could see its surface was a murky grayish brown, as dark as an asphalt road surface.

In fact, the Moon reflects only just over one-tenth of the sunlight falling onto its surface. In comparison, the Earth's reflectivity (technically known as albedo) is three times higher, while that of Saturn's moon Enceladus is almost 100 percent. If our companion was as shiny as Enceladus, we would be dazzled with a Moon almost 10 times as bright.

*The Moon's surface is pockmarked with craters of all sizes — a legacy of violent bombardment in the past.*

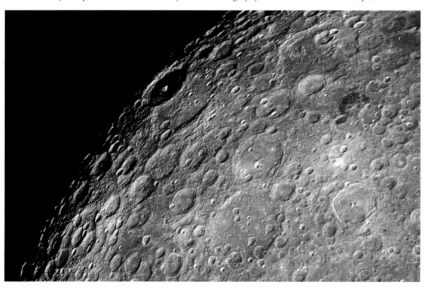

# PHASES

According to the Inuit people of Greenland, the Moon is a god, Anningan, who is forever chasing the female Sun around the sky. Anningan is so ardent that he forgets to eat, and grows thinner and thinner. Eventually, he has to come down to Earth to hunt, and the Moon disappears from the sky for three days. When he returns to the heavens, we can see him grow fatter and fatter again.

It is a colorful way of explaining the Moon's ever-changing shape, ranging from a thin crescent to a magnificent bright, round Full Moon, but the truth is much more prosaic. The Moon does not have its own light; we see it illuminated by the Sun. As the Moon travels around the Earth, we first see just a sliver of its surface lit up, then more and more until the Moon is opposite to the Sun and we can view the whole hemisphere facing us in the Sun's glare. Then, we see the illuminated portion shrink until the Moon is so close to the Sun that it is not visible at all.

*The Moon's phases, from New to Full to New again. The process takes a month — which used to be called a "moonth."*

*As the Moon circles Earth, our perspective on its sunlit half changes, creating the waxing (growing) and waning (shrinking) phases. Counterclockwise (l–r): New Moon; Crescent; First Quarter; Gibbous; Full Moon; Gibbous; Last Quarter; Crescent.*

Before the introduction of artificial lights, the Moon was essential for all nocturnal activities, especially in winter, when days are at their shortest. It was so important that people divided up the year by the coming and going of our natural night-time lantern in the sky.

And the Moon and its phases figure prominently in past literature. William Shakespeare has Juliet saying to Romeo: "O, swear not by the moon, the fickle moon, the inconstant moon, that monthly changes in her circle orb, Lest that thy love prove likewise variable."

## THE MOON: VITAL STATISTICS

| | |
|---|---|
| Distance from earth | 384,400 km |
| Orbital period (month) | 27.3 days |
| Diameter | 3,474 km |
| Mass | .012 Earths |
| Day length | 29.5 days |
| Temperature range | -330 to +250°F (-200 to +120°C) |

# THE MOON'S ORBIT

The Earth and Moon are locked in an eternal gravitational embrace, tangoing around one another in just over 27 days. Amateur astronomers can help scientists to pin down the Moon's precise path by timing phenomena called *occultations,* — the exact moment when the Moon covers up stars in its path.

The Moon's gravity stretches the Earth slightly, making the oceans move up and down as the twice-daily tides, which can reach a height of 16 m where the water is forced to pile up in the Bay of Fundy, in Canada. (The solid ground also has tides, too, but they are too small for us to notice.)

The Earth's gravity has stretched the solid Moon into a slight egg shape, with one end permanently pointing toward our planet. As a result, we only see one half of the Moon. Well,that is not *exactly* true. The Moon is slightly tipped over and its orbit is not a precise circle. As a result of these variations, which are called

## FAR SIDE

Around 41 percent of the Moon's surface is never visible from Earth. It was completely unknown until 1959, when the Russian space probe Luna 3 flew past the Moon and sent back the first photographs. The far side is heavily cratered and lacks the dark lava plains on the near side. Apollo 8 astronaut Bill Anders, a member of the first crew to see the far side, said, "The backside looks like a sand pile my kids have played in... It's all beat up, no definition, just a lot of bumps and holes."

*librations*, we can sometimes see slightly over, under or around to the far side, revealing, at different times, 59 percent of the Moon's surface.

When the Moon is closest to the Earth in its oval orbit, it appears 14 percent larger than its size at the farthest point.

*The Moon's gravity raises the tides. Here, the Moon is surrounded by a lunar halo, caused by ice crystals in Earth's upper atmosphere.*

Once a year or so, when the Full Moon occurs at its nearest point, we are treated to the brilliant phenomenon of a "supermoon."

# MOON MAP

The Moon is the most fascinating place to start exploring the Solar System. With the naked eye, you can make out the intriguing dark patches of the Man in the Moon's face. Binoculars reveal its huge mountains and even the smallest telescope makes you feel you are flying over the Moon's cratered surface.

It is natural to think that Full Moon is the best time to observe our neighbor world, but it is not. Sunlight is then illuminating the Moon evenly, and there is little contrast. It is better to check out the Moon bit by bit, at different phases, concentrating on those regions that are near the *terminator*, the line separating the bright and dark regions, where dark shadows accentuate the relief.

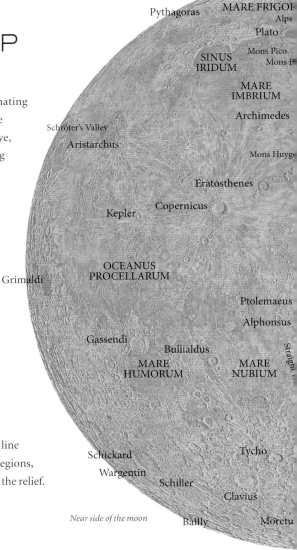

*Near side of the moon*

Aristoteles
ne Valley   Atlas
Eudoxus

Posidonius

Cleomedes

MARE
SERENITATIS

Apennines

MARE
CRISIUM

MARE
VAPORUM

ginus Rille   MARE
TRANQUILLITATIS

SINUS
MEDII

Messier

MARE
FOECUNDITATIS

Albategnius   Theophilus   Langrenus

MARE
NECTARIS

zarchel

erner

rner   Fracastorius

Petavius

Piccolomini

Maurolycus

Pitiscus

## MOON ILLUSION

The Moon looks huge in the sky, especially when it is near the horizon. But this is an illusion; the Moon is actually surprisingly small. Hold a pencil at arm's length, and it will blot out the Moon. The Full Moon looks big when it is near the horizon because we subconsciously compare it to distant objects on the skyline. You can counter this illusion effectively, if awkwardly, by turning your back to the Moon, bending over and looking between your legs: the Moon will suddenly shrink.

# LUNAR "SEAS"

Glance up at the Full Moon, and what do you see? People in many countries would probably see a round shining face, with dark eyes and a rather lop-sided mouth: the familiar Man in the Moon. But around the world, people have seen many different figures in the Moon's dark blotches. The most obvious is the Hare in the Moon (look out for the upright ears at moonrise). Across ancient Europe, people made out the shape of a man carrying a bundle of twigs, while the Chinese point out a three-legged toad.

In reality, the Moon's dark markings are huge plains of solidified lava. When the

*The Moon's seas are lava plains, created by impacts around four billion years ago.*

early Italian astronomer Giovanni Battista Riccioli viewed them with the newly

## LARGEST LUNAR MARIA

| NAME | MEANING | DIAMETER | BEST SEEN |
|------|---------|----------|-----------|
| Oceanus Procellarum | Ocean of Storms | 2,568 km | 2 days before full moon |
| Mare Frigoris | Sea of Cold | 1,596 km | first quarter |
| Mare Imbrium | Sea of Showers | 1,123 km | 1 day after first quarter |
| Mare Foecunditatis | Sea of Fecundity | 909 km | 3 days after new moon |
| Mare Tranquillitatis | Sea of Tranquility | 873 km | 2 days before first quarter |
| Mare Nubium | Sea of Clouds | 715 km | 1 day after first quarter |

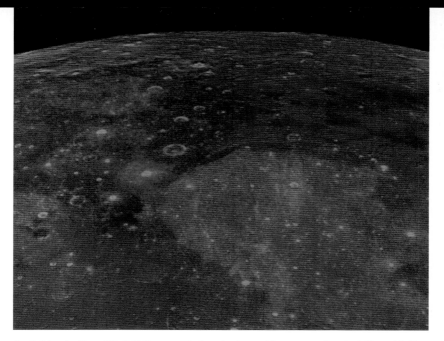

*Seas in false color: Tranquillitatis (left) appears blue from titanium enrichment; orange Serenitatis (lower right) has less titanium.*

invented telescope, he thought the smooth dark areas were great expanses of water, and gave them appropriate names based on the Latin word for "sea," *mare* (plural *maria*).

Mare Imbrium (the Sea of Showers) forms one of the eyes of the Man in the Moon, while the other eye comprises two neighboring plains, Mare Serenitatis (the Sea of Serenity) and Mare Tranquillitatis (the Sea of Tranquility). His lop-sided mouth is demarked by Mare Nubium (the Sea of Clouds) and Mare Humorum (the Sea of Moisture).

A pair of binoculars gives you a fine view of the lunar seas. Look out for the tiny but perfectly formed Mare Crisium (Sea of Crises) on the side of the Moon that the Sun first illuminates after New Moon, and, on the opposite side, the vast Oceanus Procellarum (Ocean of Storms).

# MOUNTAINS OF THE MOON

On the nights after First Quarter, you will see the Sun gradually illuminating a bright rim around one of the Moon's dark eyes. Binoculars will reveal why: a series of giant mountain chains soars above the gray lava plain of Mare Imbrium. The lunar Apennine Mountains run north-south between Mare Imbrium and Mare Serenitatis, and contain the Moon's highest peaks.

To the north, you will find the lunar Alps. Like its namesake on Earth, the range's highest peak is called (in Latin) Mons Blanc. Here, a telescope reveals the strange Alpine Valley: a long cleft through the mountains.

## FORMATION OF SEAS AND MOUNTAIN

Soon after the Moon was formed, giant asteroids smashed into its surface, gouging out vast craters called *impact basins*. The biggest is the South-Pole-Aitken basin, on the Moon's far side, which is 2,500 km across.

The Moon's near side had a thinner crust, and was warmed by an abundance of radioactive elements. As a result, the impacting asteroids melted the rocks underneath; these flowed upwards to fill the basins with lava, thus creating the dark plains of the maria.

While Earth's mountains are formed by volcanic eruptions or the planet's crust crumpling, the mountains on the Moon are simply the rims of the ancient impact basins. So we find them around the edges of the maria, especially Mare Imbrium, the biggest asteroid strike on the Moon's near side.

*The crater Plato dominates this image. To its right stretch the Lunar Alps, gashed by the Alpine Valley. Isolated peaks dot the surface of Mare Imbrium.*

Nearby, a few isolated mountain peaks, like Mons Pico, rise from the floor of Mare Imbrium; they look spectacular through a telescope as they catch the rising Sun.

## MOST PROMINENT LUNAR MOUNTAINS

| NAME | NAMED AFTER | LOCATION | HEIGHT |
| --- | --- | --- | --- |
| Mons Huygens | Christian Huygens, astronomer | Apennines | 4,700 m |
| Mons Hadley | John Hadley, instrument maker | Apennines | 4,600 m |
| Mons Bradley | James Bradley, astronomer | Apennines | 4,200 m |
| Mons Blanc | Mont Blanc, France/Italy | Alps | 3,600 m |
| Mons Pico | "Peak" in Spanish | Mare Imbrium | 2,400 m |

# CRATERS

When Galileo turned his early telescope to the Moon in 1609, he was amazed to discover it covered in round hollows. He called them "craters," after the Greek word for bowl, or cup.

In 1651, Giovanni Battista Riccioli named 247 of the most prominent craters after eminent scientists and philosophers, including himself. With increasingly powerful telescopes, and then spacecraft cameras, astronomers have increased the crater count to over a million, more than 1,500 of which have been given names. As well as great explorers like Marco Polo, and the scientists Darwin and Einstein, we find craters commemorating obscure amateur astronomers, including a Slovak priest named Hell.

## ORIGIN OF THE CRATERS

It took centuries for astronomers to work out how the Moon's craters were formed; for a long time they were thought to be giant volcanoes. But we now know the lunar craters were excavated by the impact of giant rocks from space.

The biggest craters were blasted out early in the history of the Solar System, around 3.8 billion years ago, when all the planets were peppered by wayward asteroids and comets, in an event called the Late Heavy Bombardment.

The Earth also suffered, but on our planet the ancient craters have been eroded away by rivers, glaciers and the shifting continents. The dead, airless Moon is a museum of the primaeval impacts suffered by all the planets in their infancy, with its bright *highlands* littered with craters of all sizes.

In contrast, the Moon's low-lying regions, the maria (see page 75), filled up with lava after the Late Heavy Bombardment, and these smooth plains display fewer, and generally smaller, craters. Because they appear in splendid isolation, though, craters in the maria are often more impressive, like magnificent

*The young crater Tycho, 100 million years old, is surrounded by bright rays — material ejected by the impact.*

Copernicus in Oceanus Procellarum, which was blasted out a "mere" 800 million years ago.

### SIMPLE CRATERS AND COMPLEX CRATERS

The Moon's smallest craters are plain bowl shapes called, appropriately enough, *simple craters*. Sometimes, though, even the simplest craters can have a showy side: as impacts blasted them out, streamers of light-colored rock were flung far across the Moon. You can see some of these *rays* even with the naked eye at Full Moon, stretching out from young craters like brilliant Aristarchus.

However, it gets more complicated when a bigger asteroid blows a hole over 15 km across in the Moon. The messy explosion results in a *complex crater*, each of which is so individual that lunar specialists can identify them at a glance.

First, the rocks at the center of the crater rebound upwards, to form a central mountain peak, as we find in Tycho.

The massive walls are too heavy to bear their own weight, and slump inwards to form a set of terraces, seen brilliantly in Copernicus. Finally, molten rock from the impact fills the crater to give it a wide flat floor. Sometimes, lava from an adjacent mare pushes the floor upwards, creating

a series of cracks (as in Gassendi) or, as in the case of Plato, paving the crater with a dark floor.

*Almost 100 km across, the striking crater Daedalus lies on the Moon's far side. Never visible from Earth, it's photographed here by orbiting astronauts.*

## TOP TEN CRATERS VISIBLE FROM EARTH

| NAME | NAMED AFTER | DIAMETER | BEST SEEN | NOTES |
|------|-------------|----------|-----------|-------|
| Bailly | Jean Sylvain Bailly, French astronomer | 287 km | Full Moon | On edge of Moon's disc |
| Clavius | Christopher Clavius, German astronomer | 245 km | 2 days after First Quarter | Smaller craters inside |
| Schiller | Friedrich Schiller, German philosopher | 180 km | 4 days after First Quarter | Very oval-shaped |
| Plato | Plato, Greek philosopher | 109 km | 1 day after First Quarter | Unusual dark lava floor |
| Copernicus | Nicolaus Copernicus, Polish astronomer | 93 km | 2 days after First Quarter | Surrounded by debris |
| Tycho | Tycho Brahe, Danish astronomer | 102 km | 1 day after First Quarter | Center of long ray system |
| Gassendi | Pierre Gassendi, French philosopher | 101 km | 4 days after First Quarter | Floor has giant cracks |
| Wargentin | Pehr Wargentin, Swedish astronomer | 84 km | 4 days after First Quarter | Totally filled with lava |
| Aristarchus | Aristarchus, Greek astronomer | 40 km | 4 days after First Quarter | Very bright |
| Messier | Charles Messier, French astronomer | 11 km | 3 days after Full Moon | Long comet-shaped ray |

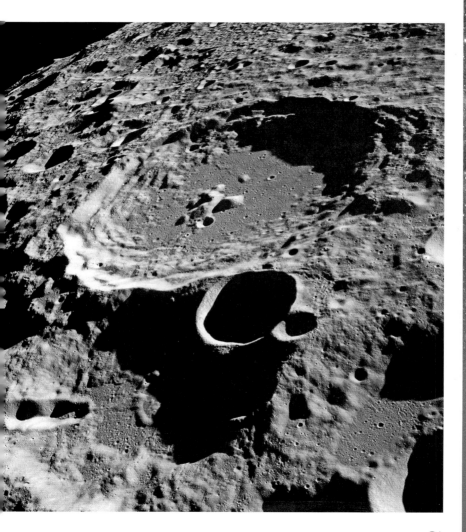

# LAVA AND VOLCANOES

I n close-up, the Moon reveals many spectacles created by volcanic forces.

### WRINKLES AND WALLS

Even binoculars will show varying shades of darkness in the maria. These differences occur because different lava flows have filled them. Mare Serenitatis resembles a target, with a lighter center surrounded by a darker ring of lava.

A telescope reveals that the mare surfaces are far from smooth; check where the shadows are long, accentuating the highs and lows. In places, where molten rock was overlying a rugged surface, it cooled unevenly to create elongated

*Schröter's Valley snakes its way from the Cobra's Head across the Moon's surface. It was cut by molten lava.*

wrinkles and these features often reveal the ghostly circular outline of a crater buried underneath.

If the light is just right (a day after First Quarter), search for the Straight Wall in Mare Nubium. It is an escarpment over 100 km long and looks like a giant cliff: in fact, it is only 300 m high, with a gentle slope. Here, the weight of lava has forced a block of the surface downwards.

## LAVA CHANNELS

Near the bright crater Aristarchus, a telescope reveals a twisting valley, flowing downwards from a depression called the Cobra's Head. Schröter's Valley, which is named after its discoverer, is a lava channel, where molten rock once flowed from the Moon's interior onto the surrounding plains.

Astronomers believe that a strange row of pits near the crater Hyginus is also the work of lava. Here, molten rock flowed through an underground lava tube, the roof of which has since collapsed in places to create the Hyginus Rille.

## VOLCANOES

If you have access to a moderately powerful telescope, scan Oceanus Procellarum near to the giant crater Copernicus, and you may pick out some small domes. These are lunar volcanoes. They are tiny compared to Earth's massive smoking mountains, less than a quarter of a mile high.

## TRANSIENT LUNAR PHENOMENA

Some astronomers claim that they have seen bright eruptions on the Moon, often near the old lava eruption site of Schröter's Valley, suggesting that our neighbor is still volcanically active. But there are no hard-and-fast scientific measurements to back up the claims. Astronomers have, however, videoed bright flashes caused by meteors hitting the Moon.

# MAN ON THE MOON

"Magnificent sight out here," commented Neil Armstrong. "Magnificent *desolation*," added Buzz Aldrin as he joined Armstrong on the lunar surface. These awed comments, from the first humans to step foot on another world, sum up the barren, rock-strewn and alien landscape of the Moon.

Between 1969 and 1972, 12 astronauts walked, hopped, drove and hit a golf ball "for miles and miles" on the Moon. These Apollo missions were the culmination of the 1960s Space Race, with the Americans determined to reach the Moon before the Russians. The astronauts were largely test pilots; just one scientist, Jack Schmitt, flew on the very last mission.

Each mission carried a set of instruments that measured moonquakes, sniffed out any possible lunar atmosphere and checked the exact distance to Earth, finding out in the process that the Moon moves 1.5 inches (4 cm) away from us every year.

Most important, the astronauts came back with one-third of a ton of Moon rocks.

## BIRTH OF THE MOON

By analyzing the lunar rocks brought back by the Apollo astronauts, scientists have made an astonishing discovery about how the Moon was born. Our companion world was created in a cataclysm that almost destroyed the Earth. Over four billion years ago, a wayward planet as big as Mars smashed into our world. The collision practically broke the Earth apart: but it pulled together again as globe of molten lava.

Incandescent drops splashed into space, to form a glowing band (like a fiery version of Saturn's rings). These droplets then came together to make up our Moon; its present serene face conceals its violent past.

*Buzz Aldrin stands by a seismic experiment deployed by the astronauts on the first manned lunar mission, Apollo 11. The Eagle lander is in the background.*

# LUNAR ECLIPSES

According to the Hupa people of California, the Moon is a great hunter, bringing home deer for his pet lions and rattlesnakes. However, sometimes they are not satisfied; they eat the Moon, leaving only a pool of blood. The Moon's 20 wives chase away the pets and sweep up the blood, and the Moon recovers.

This myth clearly refers to a total eclipse of the Moon, when our celestial companion moves into the Earth's shadow. Because the Moon's orbit is tilted, it usually doesn't pass directly behind the Earth. However, twice a year there are *eclipse seasons*, when the Full Moon strays into the shadow of the Earth.

With the illuminating sunshine cut off, we see a dark portion taken out of the Moon. Often, the eclipse is only partial, but the Moon may stray completely into the Earth's shadow and darken all over, in a total eclipse.

Look carefully, though, and you will see that the totally eclipsed Moon is not completely dark: it glows with a dull reddish color, as the Hupa so graphically described. That is because the Earth's atmosphere bends some rays of sunlight around into our planet's shadow, throwing a little light onto the Moon, even in the middle of an eclipse.

Light that travels obliquely through our atmosphere is reddened (as we see at sunset), so the light falling on the Full Moon during an eclipse has a distinctly bloody tinge.

*There is a list of forthcoming lunar eclipses, for different parts of the world, on pages 388-9.*

*The shadow of the Earth creeps over the Moon in this image of a lunar eclipse.*

# CHAPTER 4
# PLANETS

# INTRODUCTION

The Solar System is our family of fellow worlds in the cosmos. The eight planets of our local neighborhood are held by the Sun's mighty gravity, circling it with speeds ranging from dizzyingly fast to lethargic sluggishness, depending on the distance of the planet from the Sun. The innermost planet, tiny Mercury, is hurled around our local star in just 88 days. Remote Neptune, on the other hand, trundles around once every 165 years.

Seven out of the eight planets, from Mercury to Uranus, are visible to the unaided eye (but you'll need a really dark sky to spot Uranus). It is easy to work out the difference between a star and a planet. Planets, being nearby, move against the background of stars. And, unless they are very low on the horizon, they don't twinkle. That is because they appear as small discs, so that when our turbulent atmosphere washes in front of a planet, it shines steadily, unlike the distant stars, which appear as points of light and twinkle.

Many people seem amazed that you can actually *see* planets with the unaided eye, but they are among the brightest objects in the night sky, because they reflect the Sun's brilliant light. Venus is the most luminous object after the Sun and Moon.

The Solar System is also littered with debris left over from our family's birth. Millions of tiny mini-planets make up the Asteroid Belt, between Mars and Jupiter. Farther out, beyond Neptune, the Kuiper Belt is home to a myriad of larger bodies (including the now-demoted, ex-planet Pluto).

And at the very edge of the Solar System, there is a cocoon of comets: the Oort Cloud, which, jolted by the pull of a passing star, can suddenly catapult a comet toward the inner Solar System. That is when we get sensational and unexpected views of these cosmic

*Against the backdrop of the Milky Way, the eight planets of the Solar System orbit our local star, the Sun.*

snowballs, with their glorious, glowing tails boiling off in the heat of the Sun.

# SUN'S FAMILY OF PLANETS

The eight planets of our Solar System are a diverse bunch. The four worlds closest to the Sun (Mercury, Venus, Earth and Mars) are relatively small and rocky. They are all heavily cratered: a legacy of the bombardment by asteroids soon after our Solar System was born.

### ROCKY WORLDS

All of the innermost worlds have shallow atmospheres (which range from the almost non-existent on Mercury, to the thick, choking clouds of Venus). They are made of rock, with a metallic core at the center. Two of the four inner planets are accompanied by moons: Earth has one, while Mars is accompanied by two small satellites.

### GAS GIANTS

In contrast, the outermost planets (Jupiter, Saturn, Uranus and Neptune) are distended bags of gas. These worlds have no solid surface. At their heart, they may harbor a small rocky core, similar in size to the innermost planets, but they are largely made up of helium, methane and water.

And each of the giant planets lives

in a swarm of moons, buzzing around their parent world like bees. Jupiter, the mightiest planet in the Solar System, clings onto at least 67; Saturn holds court to 62 or more; Uranus has 27 moons in tow; while Neptune boasts a family of 14.

These satellite systems tumble and collide, creating a junkyard of debris in orbit around the outer planets. This debris has evolved into the beautiful ring systems we see around the gas giants today.

*The planets to scale — four inner "terrestrial" worlds and four outer "gas giants."*

# MERCURY

Look out carefully for hard-to-spot Mercury. Rumour has it that the architect of our Solar System, Nicolaus Copernicus, never observed the tiny world because of mists rising from the nearby River Vistula in Poland. Being the closest planet to the Sun, Mercury seldom strays far from the glare of our local star.

At only one-third of Earth's distance from the Sun, Mercury's parched surface is baked to 840°F (450°C) at noon and drops to -290°F (-180°C) at night. The planet rotates very slowly, compared to the time it takes to orbit the Sun. As a result, the "day" on Mercury, from noon to noon, is equivalent to 58 Earth days, while its "year" is only 88 days. On this planet, you would celebrate your birthday twice each day.

The pioneering space probe Mariner 10 sent back snatched images as it swung past the diminutive, cratered planet in 1974. It found that Mercury has a wrinkled, gnarled surface, like the skin of a dried-out apple. Astronomers believe that, as the planet cooled down, it shrank, and this caused its crust to crumple up.

And this mini-world, just one-third of the Earth's diameter across, is peppered in craters, a legacy from the millions of years of cosmic bombardment that followed the formation of the Solar System. Its biggest scar is the enormous Caloris Basin, 1,300 km wide, smashed out by the impact of a giant asteroid.

In 2011, data started to flood in when NASA's Messenger probe went into orbit around Mercury, sending back reams of data. Messenger (**Me**rcury **S**urface **S**pace **En**vironment **Ge**ochemistry and **R**anging mission) commemorates the belief in Roman mythology that fleet-footed Mercury was messenger to the gods. Messenger has found that this small world has a huge core made of molten iron. The probe has seen evidence for past volcanic activity, as well as water in Mercury's thin atmosphere.

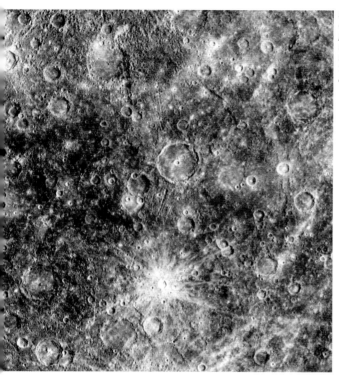

*Mercury, imaged by NASA's orbiting Messenger space probe. The planet has a negligible atmosphere, and is heavily cratered as a result. This photograph covers an area of about 1,000 km².*

Most exciting of all, peering inside permanently shaded craters at the planet's north pole, Messenger has found signs of organic compounds, and ice sheets up to 20 m thick.

# OBSERVING MERCURY

Mercury is extremely elusive. Although, at around the same brightness as the stars in Orion's belt, the planet is not faint — it can only be seen when it is near the horizon, because the diminutive world is so close to the Sun. You will only spot it at dusk and dawn. Mercury puts in three evening appearances a year, plus some early morning shows. The best times to see this little world are near the equinoxes (March and September), when it rises highest in the sky.

To find where Mercury lies in the sky, see page 384 — or consult an astronomy app or an annual stargazing book.

Don't expect to see much through binoculars. Mercury is hardly larger than our Moon and much more distant, so it looks no bigger than a point. If you have a moderate telescope, you are in better luck. You can observe the changing phases of the planet, from crescent to full and back again, as it orbits the Sun.

## TRANSITS

Mercury, being the innermost planet, can cross the face of the Sun as seen from Earth. The next "transits" are due in 2016 and 2019, when it will appear as a tiny black dot against the glowing surface of our local star. But you must NOT observe the Sun directly, because its heat could blind you; use one of the safe methods described on page 188.

## MERCURY FACTS

| | |
|---|---|
| Distance from Sun | 58 million km |
| Year | 88 days |
| Day | 59 days (rotation period) 176 days (noon to noon) |
| Tilt of axis | 0° |
| Diameter | 4,878 km |
| Mass | 0.055 Earths |
| Density | 5.4 (relative to water) |
| Gravity | 0.38 (relative to Earth) |
| Temperature | -490 to +840°F (-180 to +450°C) |
| Number of moons | 0 |

*Mercury, as seen from Earth. This amateur image required 2,815 frames stacked together to show any detail!*

The fun thing about Mercury is that it is the fastest-moving planet in the Solar System. If you have a good horizon, watch after sunset (or before dawn), and you will see it shift position from night to night.

# VENUS

Venus, named after the Roman goddess of love, is resplendent in our skies. The brightest object in the heavens, after the Sun and Moon, this world has often been mistaken for a UFO. So brilliant and beautiful, Venus can even cast a shadow in a really dark transparent sky. Its lantern-like luminosity is beguiling, but looks are deceptive.

Earth's twin in size, Venus could hardly be more different from our warm, wet world. The reason for the planet's brilliance is the highly reflective clouds that cloak its surface; probe under these palls of sulphuric acid, and you find a planet out of Hell.

Volcanoes are to blame. They have created a runaway greenhouse effect that has made Venus the hottest and most poisonous planet in the Solar System. At 860°F (460°C), this world is hotter than an oven. It boasts a thick choking atmosphere of carbon dioxide, spiked with sulphuric acid, and a surface pressure of 90 Earth atmospheres. So, if you visited Venus, you would be simultaneously roasted, crushed, corroded and suffocated.

The Soviet Space Agency sent many (unmanned!) spacecraft to land on Venus, most of which were squashed by the pressure. However, they triumphed in the end, and obtained unique pictures of Venus' superheated rocky surface.

NASA's Magellan probe, which orbited the planet from 1990 to 1994, mapped 98 percent of the planet's surface. It peered under the thick cloud using the equivalent of airport radar, revealing a world of volcanic plains, peppered with highlands. Radar also shows that Venus rotates the opposite way to the other planets; it is thought that it might have been struck obliquely by another planet in its infancy.

Why did Venus and the Earth develop so differently? Venus is closer to the Sun, just 72 percent as far as our planet, and

*Magellan's radar imaged these lava plains cloaking Venus' Eistia region. The volcano (left) is Gula Mons, 3 km high.*

that crucial input of extra heat tipped the
balance. The greenhouse effect ran wild,
and the planet boiled dry, destroying the
chance of any life surviving there.

# OBSERVING VENUS

Observing Venus is not hard. Look in the west at night, or the east in the morning (being close to our local star the Sun, Venus always follows it around the sky). It is dazzlingly brilliant, and a beautiful sight to the unaided eye.

To find where Venus lies in the sky, check out page 385, consult an astronomy app or an annual stargazing book.

Binoculars will not give you much more insight into our neighboring world (it is the closest planet to Earth), but a telescope will. As Venus circles the Sun, it shows phases, like the Moon. At some points in its orbit, it is feebly lit up as a crescent, then it is half lit and finally you get the full disc. But do not expect to see a lot through your telescope, no matter how big it is. Because the planet is covered with dense clouds, you can never see its surface. However, you may detect faint markings in its cloud patterns.

Like Mercury, Venus also transits the Sun, appearing as a black dot against our

*Amateur image of the Moon and Venus in a twilight sky, taken from Buckinghamshire, UK*

## VENUS FACTS

| | |
|---|---|
| Distance from Sun | 108.2 million km |
| Year | 225 days |
| Day | 243 days (rotation period, E to W) 117 days (noon to noon) |
| Tilt of axis | 3° |
| Diameter | 12,103 km |
| Mass | 0.81 Earths |
| Density | 5.2 (relative to water) |
| Gravity | 0.9 (relative to Earth) |
| Temperature | 864°F (462°C) |
| Number of moons | 0 |

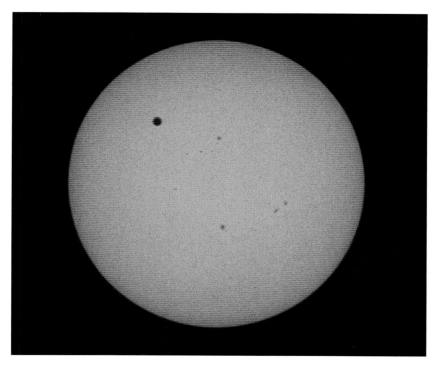

*Captured by the International Space Station, Venus transits the Sun in June 2012. Its distinct black disc contrasts with the fainter sunspots.*

local star's churning atmosphere. In 1769, expeditions from Europe traveled to the far corners of the globe to observe a transit of Venus. The goal was to calculate how far the Sun lies from the Earth, by measuring Venus' progress across our local star.

The result calculated was 153 million kilometres (only out by 3 million; the presently accepted value is 149.6 million km). There were transits in 2004 and 2012, but the next will not be until 2117.

# EARTH

"A sparkling blue and white jewel; a light, delicate sky-blue sphere laced with slowly swirling veils of white… like a small pearl in a thick sea of black mystery." What amazing world is astronaut Ed Mitchell describing here? He continues: "It takes a moment to fully realize this is Earth — home."

Mitchell was one of the few humans to have viewed our planet from as far away as the Moon. Moreover, of all the planets in our Solar System, the Earth stands apart. Warmed gently by the Sun, so it is neither perpetually baked nor permanently frozen solid, the third planet is the only one that has oceans of liquid water on its surface, providing its unique blue color.

Powered by the fiery energy from its hot core, the Earth is also an active planet; its surface is constantly shifting. Volcanoes erupt, and the ground is convulsed by powerful earthquakes.

However, what makes Earth truly unique is that it is the habitat of life. Plants have created its unusual atmosphere, rich in reactive oxygen gas, and evolution has produced something that is known only — so far — on this one planet in the entire Universe: intelligent beings.

## OBSERVING THE EARTH

You can observe our planet just by looking down at your feet, of course, while astronauts are privileged to see it as a world in space, like the other seven planets.

| EARTH FACTS | |
|---|---|
| Distance from Sun | 149.6 million km |
| Year | 365 days |
| Day | 24 hours |
| Tilt of axis | 23° |
| Diameter | 12,756 km |
| Mass | 1 Earth |
| Density | 5.5 (relative to water) |
| Gravity | 1 (relative to Earth) |
| Temperature | -135.4 to +160°F (-93 to +71°C) |
| Number of moons | 1 |

*Earthrise over the Moon, captured by William Anders on 1968's Apollo 8 around-the-Moon mission.*

However, next time you see pictures from weather satellites or spectacular images from astronauts looking down from the International Space Station, think of these as views from a visiting space probe, like the detailed images Cassini is sending back from Saturn. When we do this, we can begin to realize — like Ed Mitchell — just how beautiful, wonderful and unusual our planet really is.

# MARS

Of all the planets, Mars is the most enigmatic. For centuries it has been associated with the idea of alien life. The Italian astronomer Giovanni Schiaparelli mapped the world in 1877, when it was at its closest to the Earth. Through his telescope, he saw long, straight lines on the Martian plains, which he called "canali" (meaning "channels"). He assumed that they were natural water features. But word spread to the United States, where Percival Lowell, a super-rich Boston businessman, got the idea that they were actually "canals," constructed by intelligent Martians; he even built an observatory in Flagstaff, Arizona, to study our neighboring red world.

## MARS' MOONS

Mars is attended by two tiny moons, which are almost certainly asteroids captured by the Red Planet's gravity. Because Mars is named after the Roman god of war, its moons' names reflect the violence of conflict. Phobos ("fear" in Greek) and Deimos ("panic") are both potato-shaped and pockmarked with craters. Phobos is about 27 km long, and Deimos just half that size. Phobos is in a very low orbit about the Red Planet: it will crash into Mars in some 50 million years' time, blasting out a crater 300 km across.

Mars is much colder than the Earth, as it orbits 52 percent farther from the Sun, and it is also smaller. But the Red Planet shares much with our planet. It has an atmosphere (albeit very thin), polar caps and sensational geology. The planet boasts a canyon, Valles Marineris, that is 4,000 km long and 7 km deep. It also has an astounding network of volcanoes, some of which might be poised to become active. The biggest is Olympus Mons. It is three times higher than Mount Everest, and big enough to cover the whole of Spain.

Mars' redness is a result of it being rusty. Early on, water from its soils reacted with the iron in the planet's surface, leading to its ruddy color. Sometimes the red is interrupted by a dust storm, which obscures the planet's features. But of all the other planets in the Solar System, it is the most active and inviting world.

*Bluish-white clouds of ice crystals hang above the Tharsis region of volcanoes on Mars. This image was captured in April 1999 by the Mars Orbiter Camera, operating on the Mars Global Surveyor spacecraft.*

# LIFE ON MARS

On October 30, 1938, 23-year-old Orson Welles sent terror into the heart of America, a terror that even made people flee their homes. Purporting to be a continuity announcer, actor Welles interrupted a program on CBS radio with: "Ladies and gentlemen. I have a grave announcement to make. The strange object that fell at Grovers Hill, New Jersey, was not a meteorite. Incredible as it seems, it contained strange beings who are believed to be a vanguard of an army from the planet Mars."

The Washington correspondent came in with tales of brutal massacres by "the Tripods" that had landed, and total panic ensued.

But, in reality, it was an adaptation of *The War of the Worlds*, by H. G. Wells. The hoax was designed to boost the flagging ratings of CBS, and it turned Orson Welles into a legend. It also made people think about whether there *was* life on the Red Planet.

The first space probes sent to Mars, however, revealed no Tripods or Little Green Men: just a cold, barren desert.

However, in 1976, the twin NASA Viking probes landed on Mars. There were four experiments on board, designed specifically to look for traces of life. One experiment proved positive. The results were controversial, but we find in these data strong evidence for bacteria, "little green slime."

And there is more hope for finding life on our neighboring world… The present flotilla of space probes, crawling over or orbiting the Red Planet, are unanimously picking up evidence for present or past water, the essential ingredient for life, all over Mars.

NASA's Curiosity mission, which landed on Mars in August 2012, is actively sniffing around Gale Crater. Its rover, which is the size of a small car, is investigating the geology, composition and potential for life of the Red Planet. For the first time, NASA acknowledges that it has designed the project with a view toward looking at the possibility of humans going to Mars. The privately funded Mars 1 mission could deliver a crew to the Red Planet by as early as 2024. Then we will know that there *is* life on Mars.

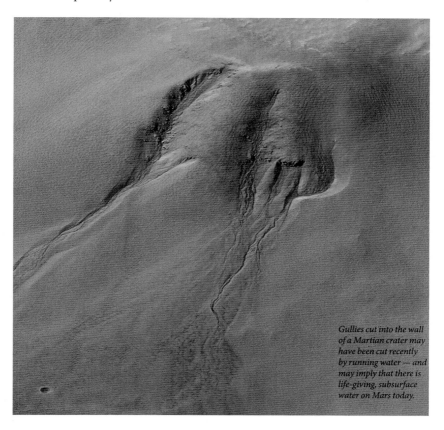

*Gullies cut into the wall of a Martian crater may have been cut recently by running water — and may imply that there is life-giving, subsurface water on Mars today.*

# MARS MAPS

For such a small planet, just half the size of Earth, Mars boasts more extreme geology than our world: deep canyons, colossal volcanoes and a dry, barren planet-wide desert. Yet space probes have evidence that water was abundant on the Red Planet in the past and may still be present under its sands today.

Here are Mars' major geological features:

## Olympus Mons

Fittingly named after Mount Olympus, this is the biggest volcano in the Solar System. At 26 km, the volcano is three times higher than Mount Everest. Although it is believed to be extinct, there is a theory that Olympus Mons might possibly erupt again.

VASTITAS BOREA

ARCADIA
PLANITIA

Alba Pa

AMAZONIS
PLANITIA

Olympus Mons

Ascraeus Mo

Pavonis Mons

THARSIS BULGE

Arsia Mons

DAEDALIA PLANUM

SOLIS
PLANUM

TERRA
SIRENUM

Newton

Copernicus

Chamberlin

AONIA
TERRA

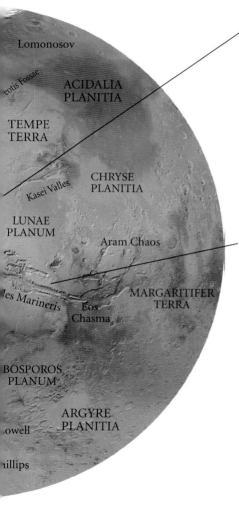

Lomonosov

eotis Fossae

ACIDALIA
PLANITIA

TEMPE
TERRA

Kasei Valles

CHRYSE
PLANITIA

LUNAE
PLANUM

Aram Chaos

es Marineris    Eos
Chasma

MARGARITIFER
TERRA

BOSPOROS
PLANUM

ARGYRE
PLANITIA

owell

illips

## THARSIS BULGE

This is Mars' major volcanic region, and
the cause of the Valles Marineris rifts. As
lava welled up from below the surface,
the planet cracked apart, like the skin
of an over-ripe tomato. Tharsis is 4,000
km across and rises 10,000 m above
the Martian plains. It boasts three huge
volcanoes, all of them bigger than any
volcanoes on Earth.

## VALLES MARINERIS

This vast gash across the face of Mars is
so huge that you could fit the whole of
the Earth's Alps inside it, and still have
room to spare. It is named after Mariner
9, which was the first probe to orbit the
planet, in 1971. The canyon stretches
4,000 km, 10 times longer than the Grand
Canyon. In places, it is 200 km wide and
7 km deep.

## SYRTIS MAJOR

This is the most prominent dark feature on the face of Mars. Once, astronomers believed that the planet's dark markings were vegetation, as they changed shape with the Martian seasons. Now we know that the changes are due to windborne dust from the deserts, which periodically cover and uncover the markings. The triangular Syrtis Major is about 1,000 km across. Space-probe observations have revealed that it is a low-relief volcano, and it owes its dark color to basaltic rock.

## HELLAS PLANITIA

This huge, circular hole in the Martian surface is named after Greece and is among the three biggest impact craters in the Solar System. It was one of the first features on the Red Planet to be observed through a telescope. The scar is 7,000 km deep and 2,300 km wide, larger than the Caribbean Sea. It was gouged out by an asteroid that hit the planet during the Late Heavy Bombardment, around four billion years ago. Gullies on the crater's surface indicate signs of glacial activity in the past.

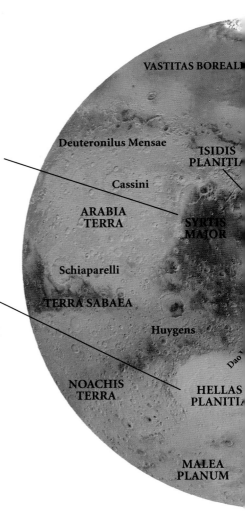

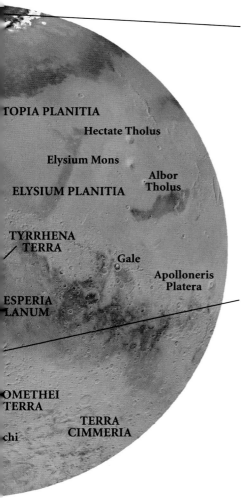

TOPIA PLANITIA

Hectate Tholus

Elysium Mons

Albor
Tholus

ELYSIUM PLANITIA

TYRRHENA
/ TERRA

Gale

Apolloneris
Platera

ESPERIA
LANUM

OMETHEI
TERRA

TERRA
CIMMERIA

chi

## POLAR CAPS

Like the Earth, Mars has two polar ice
caps. The North Pole's outranks that of the
South Pole: it comes in at 1,100 km across
while its southern counterpart is 400 km
wide. Both are made of water ice, with
an overlay of carbon dioxide ice. Mixed
into this chemical melée are layers of
Martian dust, all of which may help future
explorers to date events in the history
of the Red Planet. Every spring, the
carbon dioxide in the polar caps sublimes
(evaporates), creating ferocious winds of
up to 400 km per hour.

## RIVER CHANNELS

Think about Mars, and you are never far
from questions about water, and space
probes have discovered that Mars had a
decidedly wet past. Detailed images have
shown dried-up river channels on the Red
Planet. Dao Vallis and its tributary, Niger
Vallis, stretch for over 1,200 km. Some
of these rivers flowed just a few tens of
millions of years ago—very recently on
the astronomical timescale.

# OBSERVING MARS

Mars' "year" is 687 days, so the Earth and the Red Planet only line up every two years. When it does arrive in the sky for a few months, it can put on a glorious show. Its ruddy color outdazzles most of the stars, and, as an object visible to the unaided eye, it is outstanding.

To find where the Red Planet planet lies in the sky, check out page 385–6, or consult an astronomy app or an annual stargazing book.

However, expect to be disappointed when you try to observe our neighboring world through an optical aid. Binoculars will show nothing. Mars is too small and far away. A medium-sized telescope will reveal the most prominent dark markings, like Syrtis Major. You can track them as the planet spins (its "day" is only half an hour longer than the Earth's).

However, watch out for white-outs. Mars gets engulfed in dust storms, and its surface features disappear altogether. These are seasonal, like the hurricane and tornado outbreaks on our own planet, and they can blot out all its features.

Mars' seasons occur for the same reason as those of the Earth: the planet's axis is at an angle to the Sun, so its northern and southern hemispheres receive different amounts of sunlight as it circles our local star. The changes affect the polar caps, which swell and shrink throughout Mars' year, and are worth looking out for if you have a telescope.

## MARS FACTS

| | |
|---|---|
| Distance from Sun | 227.9 million km |
| Year | 687 days |
| Day | 24 hours 37 minutes |
| Tilt of axis | 25° |
| Diameter | 6,792 km |
| Mass | 0.11 Earths |
| Density | 3.9 (relative to water) |
| Gravity | 0.38 (relative to Earth) |
| Temperature | -225 to +95°F (-143 to +35°C) |
| Number of moons | 2 |

*Triangular Syrtis Major dominates this amateur image of Mars, with the pale round basin of Hellas above. South is at the top (usual in amateur photos through an inverting telescope).*

Can you observe the Red Planet's moons? Phobos (although bigger than Deimos) is hard to spot, because it is so close to Mars. Its little brother is a better bet, although we recommend a telescope with an aperture of 250 mm or more.

# JUPITER

Jupiter is the King of the Solar System, named after the chief Roman god. It's big enough to contain 1,300 Earths; and, as it is made almost entirely of gas, the planet is very efficient at reflecting sunlight. Despite its vast bulk, Jupiter spins faster than any other planet in the Solar System, with a day lasting less than 10 hours, which makes its equator bulge outwards.

A fearsome magnetic field creates stunning aurorae at Jupiter's poles. But no astronaut could survive the radiation, not to mention the violent lightning storms.

*Jupiter's fearsome spin stretches its clouds into streaks girdling the giant world. The Great Red Spot (bottom-left) is a huge vortex in the planet's atmosphere.*

## THE CORE

The planet's heart creates more energy than it receives from the Sun. The mysterious interior of Jupiter is largely made of hydrogen, at such great pressure that it behaves like a liquid metal, and simmers at a temperature of 63,000°F (35,000°C). Had Jupiter been around 75 times more massive, its core would have been hot enough to fuse hydrogen to form helium, the process that powers the Sun, and the planet would have become a star.

## MISSIONS TO JUPITER

Although Jupiter lies so far away (over five times the Earth's distance from the Sun), the giant planet has had many visitors from our planet, in the shape of unmanned space probes. The first were two Pioneers, followed by the twin Voyagers. Voyager 1 discovered that Jupiter is circled by three faint rings, made of dust particles that have drifted off its moons Metis and Adrastea.

The first probe to orbit Jupiter was Galileo, in 1995. En route, it recorded spectacular images of the doomed Comet Shoemaker-Levy 9 crashing into the giant planet in 1994. Galileo swung around the planet and its moons for over seven years, when it was deliberately destroyed in case it impacted with the moon Europa, which might harbor life.

Galileo dropped a probe into Jupiter's atmosphere to study its structure and composition. The probe plummeted to depths where the pressure was 22 times that of the Earth's atmosphere, and the temperature over 320°F (150°C). It was almost certainly vaporized, but the mission was successful.

# JUPITER MAP

Through a small telescope, Jupiter looks a bit like a tangerine crossed with an old-fashioned peppermint. The peppermint stripes are cloud belts of ammonia and methane stretched out by the planet's dizzy spin. Jupiter's striped disc is divided into dark belts and pale zones. The belts are where you can see more deeply into the planet's atmosphere.

There can never be an up-to-date map of the giant planet because cloud features are changing all the time, but there are several features to pick out.

## GRS: THE GREAT RED SPOT

In the late 17th century, Italian astronomer Giovanni Domenico Cassini found a dark marking on Jupiter, but it faded from view. The present Great Red Spot has been visible for about two centuries. The Great Red Spot is an enormous anticyclonic storm, lying 22 degrees south of the planet's equator. Its top hovers 8 km above Jupiter's main cloud base, making it cooler than its surroundings.

The spot spins, with a period of about six Earth days. It is powered by violent winds: those at the edge of the spot reach over 430 kph. The upward-spiralling winds carry gases to great heights in Jupiter's atmosphere, where they react with sunlight.

The enormous spot, some 40,000 km across, could swallow up three Earth-sized planets, but in recent years this much-admired feature has been shrinking. However, astronomers studying Jupiter's climate predict that it won't fade away.

Why is it red? The answer is unknown because the spot's color changes from true red, to salmon-pink to near white. The best bet is on phosphorus atoms reacting with radiation from the Sun.

*The Great Red Spot is imaged in intimate detail by the Cassini spacecraft en route to Saturn.*

GRS

### NPR: NORTH POLAR REGION

Thousands of individual storms here are racing around Jupiter's pole.

### NTB: NORTH TEMPERATE BELT

The most northerly of the major belts, it fades from view every decade or so.

### NTrZ: NORTH TROPICAL ZONE

This is the highest of the zones, lying above the belts. It's sometimes bordered by "red ovals" — eddies in the atmosphere almost half the size of Earth.

### NEB: NORTH EQUATORIAL BELT

This twisted structure is caused by violent winds. Jupiter's winds are highly complex, with adjacent currents roaring past in opposite directions.

### EZ: EQUATORIAL ZONE

A region of stable clouds whizzing around the planet's equator

### SEB: SOUTH EQUATORIAL BELT

Variable in brightness, it's the lair of the Great Red Spot.

### STrZ: SOUTH TROPICAL ZONE

Its high-altitude clouds are made of white ammonia crystals.

### SPR: SOUTH POLAR REGION

This complex area of turbulence is frequently home to "white ovals" — enormous temporary storm systems.

### OVAL BA (OUT OF VIEW)

Lying to the south of the Great Red Spot, it is testimony to the planet's volatile atmosphere. In 2000, three white oval storms began to merge. By 2005, their union was starting to turn red — possibly as a result of the effects of sunlight. Astronomers affectionately nickname it "Red Spot Junior." Will it become a permanent feature? Only Jupiter's restless atmosphere will tell.

# JUPITER'S MOONS

Jupiter commands its own "mini-solar system": a family of almost 70 moons. The four largest are visible in good binoculars and even, for the really sharp-sighted, to the unaided eye. Galileo was probably the first to observe them through the newly invented telescope, in 1610.

These are worlds in their own right; Ganymede is even bigger than the planet Mercury. But two vie for "superstar" status: the surface of Io is erupting, while brilliant white Europa has liquid water beneath its solid ice coating. The four biggest moons, all named after Jupiter's lovers, are as follows:

### IO

Making it look like a cosmic pizza, the red, orange and yellow blotches on Io (diameter 3,640 km) come from volcanic plumes shooting sulphur dioxide some 300 km into space. This moon is the most active world in our Solar System.

*Volcanic Io*

*Smooth Europa*

The active volcanoes are the result of Io being pummelled by Jupiter's mighty gravitational field, which stirs up the moon's interior.

### EUROPA

Smoother than a billiard ball, the smallest of Jupiter's four large moons (diameter 3,100 km) is the least understood. It is wrapped in a deep layer of ice — and underneath lies a huge ocean, where alien fish may swim…

### GANYMEDE

At 5,268 km across, Ganymede is the largest moon in the Solar System. Its icy crust is heavily cratered from cosmic impacts soon after it was formed. However, Ganymede's surface is also criss-crossed with a complex system of ridges and grooves, signs of more recent activity.

### CALLISTO

There are no "plains" on the surface of this 4,820 km diameter moon, which is densely packed with impact craters. Valhalla, 300 km across, is the biggest. Callisto resembles our Moon — but its craters are gouged into ice, instead of rock.

*Giant Ganymede*

*Cratered Callisto*

# OBSERVING JUPITER

Jupiter is a fantastic target for stargazers, whether you are using your unaided eye, binoculars or a small telescope. And it is surrounded by an entourage of moons, which all add to the fun.

The giant planet is the third-brightest object in the night sky, after the Moon and Venus. It hardly moves from night to night; Jupiter circles the Sun in just under 12 years, making it seem almost stationary amongst the stars. But it is a glorious sight.

To find where the giant planet lies in the sky, check out page 386 — or consult an astronomy app or an annual stargazing book.

With the unaided eye, you will see a very bright, non-twinkling "star." Like all the planets, it presents a disc to the cosmos, not just a point of light like the distant stars.

Use binoculars, and you will catch the moons Io, Europa, Ganymede and Callisto. Night by night, you will see them change position. Sometimes you can see all four moons, but in their dance around the planet they disappear and reappear behind Jupiter's great bulk.

With a telescope, you are in for a huge treat. First of all, it is amazing to look at the planet's flattened, bulging shape, the result of its crazy spin. Then there is the fantastic texture of Jupiter's belts and zones. Watch over a few hours, and you can actually see the planet rotating.

## JUPITER FACTS

| | |
|---|---|
| Distance from Sun | 778 million km |
| Year | 11.9 years |
| Day | 9 hours 55 minutes |
| Tilt of axis | 3° |
| Diameter | 142,980 km |
| Mass | 318 Earths |
| Density | 1.3 (relative to water) |
| Gravity | 2.5 (relative to Earth) |
| Temperature | -166°F (-110°C) |
| Number of moons | 67 (as of 2014) |

*Galileo's telescope log, revealing the movement of Jupiter's four biggest moons from night to night.*

With a telescope you will also be able to see more of the motions of Jupiter's moons. As they move around the giant planet, they sometimes pass in front of their massive father world, appearing as a tiny disc accompanied by a dark shadow against Jupiter's churning clouds.

Jupiter's atmosphere changes from day to day, much like the Earth's. But it is as a result of the efforts of amateur astronomers worldwide, who have logged significant changes in its cloud patterns, that we know so much about the workings of Jupiter. Out every night, they note

*Detailed amateur photograph of Jupiter. The Red Spot appears at top — telescopes invert images, and observers prefer to stay with the conventional inverted view.*

the changes on the giant planet: both disturbances in the belts and zones and new "white ovals."

Jupiter is a phenomenally active world, and an excellent excuse for buying a medium-powered telescope.

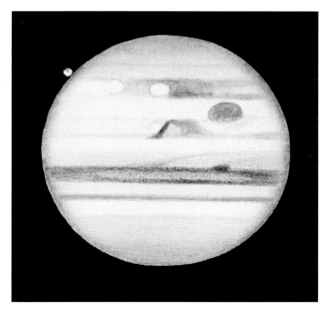

*Amateur drawing of Jupiter (also inverted). The observer was using a 25 mm reflector, and noted: "Ganymede entering occultation. Great Red spot blatantly red; other colors subtle."*

# SATURN

The slowly moving Saturn is famed for its huge ring system. Galileo, using his primitive telescope, was baffled by the rings; with the low resolution of his instrument, he thought that Saturn was a triple planet.

In 1665, Christiaan Huygens, using a telescope that he had designed with 50x magnification, discovered the true nature of the planet's "companions." "It is surrounded by a thin, flat ring, nowhere touching…" he wrote.

The ring system is the biggest in the Solar System: it would stretch nearly all the way from the Earth to the Moon. Made of chunks of ice, ranging in size from particles smaller than frozen peas to blocks as large as refrigerators, this encircling halo is almost certainly the residue of an icy moon that was broken up by Saturn's gravity.

Saturn itself is second only to Jupiter in size. But it is so low in density that, were you to plop it in an ocean, it would float.

Like Jupiter, Saturn has a ferocious spin rate, of 10 hours and 34 minutes, and its winds roar at speeds of up to 1,800 kph.

However creamy-yellow Saturn's atmosphere is much blander than that of its larger cousin, probably due to a haze of ammonia crystals in its upper atmosphere.

Underneath, Saturn shares the same banded appearance as Jupiter, although its markings are much more subtle. However, roughly every 30 years, when it reaches its "summer solstice," with the planet's northern hemisphere inclined toward the Sun, Saturn can break out in a rash of white spots.

Four space probes have crossed the immense gulf to Saturn, which orbits the Sun almost 10 times farther out than Earth, to visit the ringed world. Pioneer 11 and Voyagers 1 and 2 merely whizzed past the planet. Cassini is currently in orbit about Saturn, studying the planet's globe, rings and its 62 or more moons. Cassini has discovered that Saturn is not as staid

*Unusual angle on ochre-tinged Saturn, as the orbiting Cassini spacecraft flies over the planet*

as it looks; the atmosphere is wracked with lightning bolts 1,000 times more powerful than those on Earth.

# SATURN MAP

Any map of Saturn will be dominated by its magnificent rings. It was Giovanni Domenico Cassini who observed in 1675 that Saturn is girdled by a number of separate rings. Close-up scrutiny from space probes reveals that the rings themselves are made up of a myriad of thin ringlets, each composed of icy particles and chunks of ice.

## A RING

This is the outermost of Saturn's three bright rings. It is divided by a 300 km wide gap, the Encke Gap, which is swept clean of particles by the tiny moon Pan. For all their enormous width, the rings are paper-thin: the A ring is no more than 10 to 30 m thick.

## CASSINI DIVISION

Nearly 5,000 km across, the Cassini Division is a huge gash between rings A and B. Discovered by Cassini through his 6.1 m long telescope, it is the result of the gravity of Saturn's moon Mimas pulling on the ring particles.

## B RING

The broadest and brightest of Saturn's rings, it is also the densest. The B ring is crossed by dark "spokes" (visible only from space probes) composed of fine particles of dust channelled into lines by electric fields.

## C RING

The so-called "crepe ring" is made of darker particles than rings A and B, and is just 5 m thick.

## F RING

This ring, Saturn's outermost, was discovered by Pioneer 11 in 1979. Lying 3,000 km beyond the A ring, this is the most active in all of the planet's ring system. It is a very thin and convoluted feature, and its appearance changes from hour to hour. Its particles are "shepherded" by a small pair of moons, Prometheus and Pandora, which corral the particles in this delicate wraith of ice.

## SATURN'S GLOBE

Compared with Jupiter, Saturn is a bland world. The poles offer a little more excitement, although you would have to be in a spacecraft to see them. At the North Pole, the clouds make up a giant hexagonal pattern. At the south, there is a polar vortex, with winds raging at 550 kph.

A RING   CASSINI DIVISION   F RING

B RING   C RING                    SATURN'S GLOBE

# SATURN'S MOONS

The iconic rings are just the beginnings of Saturn's extensive family. It has 62 known moons, including giant Titan. In 2014, astronomers saw signs of another tiny moon, informally called "Peggy," being born from icy chunks at the edge of Saturn's main ring system. The international Cassini-Huygens mission has discovered that these moons are far more exciting than Earth's Moon, and some may harbor life. The following are the top five.

*Ridges and fractures on Saturn's icy moon Enceladus, captured by NASA's Cassini spacecraft in 2009*

### TITAN

Bigger than the planet Mercury, Titan is the only moon in the Solar System to boast a thick atmosphere, with a pressure 50 percent greater than that on Earth. Its "air" is mainly made up of nitrogen, like our own world, but there the similarities end. Titan is a moon in deep freeze: it struggles in temperatures of -290°F (-180°C). In 2005, Europe's Huygens probe penetrated Titan's murky orange clouds and landed on the moon's surface. It discovered seas of liquid ethane and methane, ideal environments for the development of future life.

### ENCELADUS

Enceladus is a mere 500 km across, but scientists believe that it has the greatest potential in the Solar System for harboring extraterrestrial life. There is evidence for liquid water under its icy crust, from which

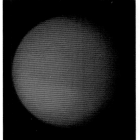

*Cloud-cloaked Titan*

*Icy Enceladus*

*Mutilated Mimas*

giant plumes spew into space. The ice particles from these erupting *cryovolcanoes* have created one of Saturn's fainter rings.

### MIMAS

Tiny Mimas, just 400 km across, was discovered by British astronomer William Herschel in 1789. Its main feature is a gigantic crater, looking like a cosmic eye, created in an impact that almost broke the moon apart. Appropriately, the crater is called Herschel.

### HYPERION

This is one of the most deformed moons found in the Solar System: an irregularly shaped body, covered in craters. Hyperion may have been born in a collision between moons. While its surface is dark and rocky, most of the moon, like the majority of Saturn's satellites, is made of ice.

### IAPETUS

This moon has always baffled astronomers. One side of it is as black as coal, while the other side is as white as freshly fallen snow. The contrast is probably caused by ice subliming from the moon's surface, and darkening as it evaporates. Debris blasted off Iapetus' neighboring moons may also add to its sultry color.

# OBSERVING SATURN

While the planets from Venus to Jupiter can outshine any of the stars, Saturn is much duller to the naked eye: a yellowish, steadily glowing point of light, a bit fainter than the most brilliant stars.

To find where Saturn lies in the sky, check out page 386–7, or consult an astronomy app or an annual stargazing book.

However, through even a small telescope Saturn is an incredible sight. It's unlike any other planet: a ringed world hanging in the blackness of space. It looks unreal, like an artwork out of a science fiction comic. Because of its rapid spin, Saturn is the most flattened planet in the solar system and its bulging equator is easy to see using a medium-sized telescope (over 150 mm across).

Do not expect to see much detail on Saturn's disc. Its features are shrouded in a layer of haze, and its cloud belts are much subtler than those of Jupiter. However, if you look at the planet's equator, you'll see a very wide band girdling its waistline.

Saturn can pull off surprises, such as the white spots that appear roughly every 30 years. The most famous case was in 1933, when the film comedian Will Hay, a passionate amateur astronomer, discovered a white spot on the usually somnolent planet. Recently, there have been several outbreaks, including a global

*A violent storm on Saturn, captured in this amateur image from Sussex, UK, in March of 2011.*

## SATURN FACTS

| | |
|---|---|
| Distance from Sun | 1,433 million km |
| Year | 29.5 years |
| Day | 10 hours 34 minutes |
| Tilt of axis | 27° |
| Diameter | 120,540 km |
| Mass | 95 Earths |
| Density | 0.7 (relative to water) |
| Gravity | 1.1 (relative to Earth) |
| Temperature | -220°F (-140°C) |
| Number of moons | 62 (as of 2014) |

storm in 2011, and you never know when Saturn's activity might pick up again.

Saturn's rings are a joy to observe. During the planet's long year, their angle to us changes. Sometimes they are in-your-face, full-on and very bright. At other times, the rings are edge-on and disappear from sight.

When Saturn is at opposition (in line with the Earth and the Sun), the icy ring particles briefly reflect the Sun's light toward us and the planet noticeably brightens. This is an effect you can see even with the naked eye.

Look through a medium-sized telescope and you can watch the movement of Saturn's biggest moons. Chief among them is mighty Titan, which you can spot with even a small telescope. Then we have Rhea, Dione, Tethys and Iapetus. However, because of the latter's piebald nature, it changes a lot in brightness. So don't be disappointed if you cannot catch Iapetus on your first attempt.

*On August 3, 1933, the respected stage and screen comedian Will Hay — also a keen amateur astronomer — discovered a huge white spot on Saturn, using a 6-inch refractor. His sketch of the "Great White Spot" reveals that the planet can become unexpectedly lively!*

# URANUS

It was an unlikely discovery. William Herschel, musician, composer and organist at the Octagon Chapel in Bath, UK, was a passionate amateur astronomer. His obsession was to build bigger and bigger telescopes to survey the sky.

One night, in 1781, he found a faint, unknown greenish blob. By the next night, it had moved against the nearby stars. Herschel thought that he had stumbled upon a comet, but the object turned out to be a planet.

Uranus was the first planet to be found since antiquity, and its discovery literally doubled the known diameter of our Solar System; it lies 19 times as far from the Sun as the Earth.

Uranus is a gas giant like Jupiter and Saturn. Four times the diameter of the Earth, it has an odd claim to fame: it orbits the Sun on its side (probably as a result of a collision in its infancy, which would have knocked it off its axis).

Like the other gas giants, it has an encircling system of rings. But these are nothing like the spectacular edifices that girdle Saturn: the 13 rings are thin and faint. Many of us were disappointed when the Voyager probe flew past Uranus in 1986 to reveal a bland, featureless world, but the planet is becoming more active as its seasons change, with streaks and clouds appearing in its atmosphere. Giant storms erupted in August 2014.

## THE MOONS OF URANUS

The planet boasts a family of at least 27 moons, named after characters in the writings of Shakespeare and Alexander Pope. The top four are as follows:

### MIRANDA

This tiny moon, less than 500 km across, looks a mess, crumpled up with grooves, craters and cliffs. It was probably smashed apart by a huge impact, but reassembled itself.

## ARIEL

The surface of this small moon is heavily grooved; some of the gashes are up to 200 km long. However, there are also plains, which may point to previous volcanic activity.

## UMBRIEL

The third-largest and darkest of Uranus' moons, Umbriel is very heavily cratered.

## TITANIA

At nearly 1,600 km in diameter, Titania is the biggest moon circling Uranus. Partly rock, partly ice, it rejoices in a large crater called Gertrude (the mother of Hamlet), which is 326 km across.

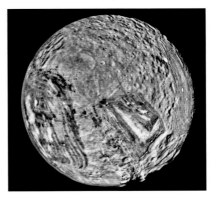

*Bland Uranus (top) and its battered moon Miranda (bottom), imaged by Voyager 2.*

# NEPTUNE

With Pluto being demoted to a mere "dwarf planet," Neptune is officially the most remote planet in our Solar System. It lies at 30 times the Earth's distance from the Sun, in the twilight zone of our family of worlds, and it takes nearly 165 years to complete one orbit.

Neptune is a planet that was discovered by the power of mathematics. After Uranus was found, astronomers realized that it was being pulled off course by an unknown gravitational force, perhaps from another planet lying farther out.

Two mathematicians, John Couch Adams in England and Urbain Leverrier in France, independently calculated where the missing planet should be. Primed by these calculations, German astronomer Johann Galle tracked down the object in 1846.

You need a space probe to get up close to this gas giant planet. In 1989, Voyager 2 unveiled a turquoise world 17 times heavier than Earth, cloaked in clouds of methane and ammonia.

For a planet so far from the Sun, Neptune is amazingly active. Its core blazes at nearly 9,000°C (5,000°F), almost as hot as the Sun's surface. This internal heat drives dramatic storm systems and dark spots, as well as winds of 2,000 kph: the fastest in the Solar System.

Like all the distant gas giants, Neptune is circled by a ring system. Although not a match for those of Saturn, it boasts three faint rings. This most distant planet also has a family of 14 moons, including Triton: the coldest object in the Solar System.

*Freezing Triton has geysers that erupt ice.*

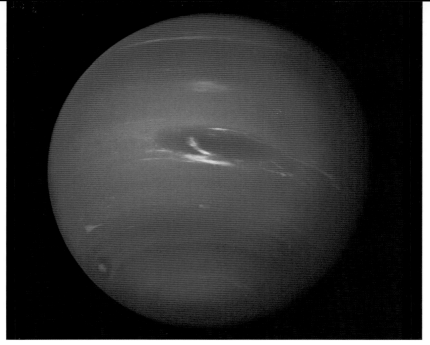

*Usually livelier than Uranus, Neptune sported a "Great Dark Spot" when Voyager 2 flew past in 1989.*

### TRITON

This moon is an enigma. It's large; at 2,700 km across, Triton is the seventh-biggest moon in the Solar System. For such a frozen world, it is amazing that Triton is geologically active. It boasts geysers that erupt plumes of nitrogen and dust into space.

Triton circles Neptune in a backwards ("retrograde") direction, which strongly suggests that it was captured from the neighboring Kuiper belt (see page 156). This may spell Triton's doom, because it is very close to its adopted planet: tidal forces and interactions with orbital debris may send it spiralling inward 3.6 billion years from now. Triton will either collide with the planet or endow it with another ring.

# OBSERVING URANUS AND NEPTUNE

If you are very sharp-sighted, and in a dark, transparent site, you can just see Uranus with the unaided eye. Binoculars are a huge help — you will be able to pick out a distinctly greenish "star" that slowly moves against the background stars. Even through a telescope, don't expect to see any detail on Uranus; it is the blandest planet of them all. Its moons are very hard to spot, too, unless you are lucky enough to have a whopping light-grabber.

Neptune is a telescope-only planet. Not visible to the unaided eye, seeing the most distant planet of the Solar System requires optical aid. However, as a bonus, a good telescope will also reveal its giant moon Triton.

How to find these remote worlds? As a general guide, we list the constellation they currently inhabit on page 387 — but you'll need more detailed directions to track them down for a specific date. Use

## URANUS FACTS

| | |
|---|---|
| Distance from Sun | 2,877 million km |
| Year | 84 years |
| Day | 17 hours 14 minutes |
| Tilt of axis | 82° |
| Diameter | 51,120 km |
| Mass | 15 Earths |
| Density | 1.3 (relative to water) |
| Gravity | 0.9 (relative to Earth) |
| Temperature | -330°C (-200°F) |
| Number of moons | 27 (as of 2014) |

## NEPTUNE FACTS

| | |
|---|---|
| Distance from Sun | 4,503 million km |
| Year | 165 years |
| Day | 16 hours 6 minutes |
| Tilt of axis | 28° |
| Diameter | 49,530 km |
| Mass | 17 Earths |
| Density | 1.6 (relative to water) |
| Gravity | 1.1 (relative to Earth) |
| Temperature | -330°C (-200°F) |
| Number of moons | 14 (as of 2014) |

*Uranus and its brightest moons — big telescopes are essential to view them!*

planetarium software to print out a map you can use at your telescope. Or if you have a GO TO (see page 49) telescope that will automatically home in on the planets' positions, use that.

It is quite awesome to look at these lonely worlds at the edge of the Solar System, and, when you find them for yourself, to feel that you're truly at the frontier.

# CHAPTER 5

# COSMIC VERMIN

# INTRODUCTION

In 1803, the inhabitants of L'Aigle, France, were astonished to see thousands of stones coming crashing down to the ground from space. People watched in disbelief and many eminent scientists did not believe them; there were planets and stars beyond the Earth, but not stones!

Perhaps, they surmised, these rocks were thrown out by a volcano, or had coalesced in the atmosphere, like hailstones. But a young physicist, Jean-Baptiste Biot, traveled to the scene. There, he interviewed the witnesses and analyzed the stones, proving that these were wayward rocks from space.

## BIRTH OF THE SOLAR SYSTEM

Around 4.6 billion years ago, the Sun and planets did not exist. Then, gravity pulled together a cloud of gas and dust in interstellar space. As it shrunk in size, the cloud spun around faster, shaping itself into a thin disc, like an Italian cook spinning a cosmic pizza.

The central region condensed into a shining ball of gas: the Sun. In the swirling disc, microscopic particles of dust stuck together, to build up into rocky pebbles. Farther from the Sun, where it was colder, ice crystals agglomerated to form cosmic snowballs.

Gravity pulled these solid chunks together. The pebbles built up the rocky planets near the Sun, while enormous numbers of snowballs formed the gas giants farther out. However, a huge amount of debris of all kinds was left over — the rubble from the construction of the planets.

*Worlds in the making: around our fledgling Sun, dust and debris crash violently together to create the planets.*

Today, we know that the Solar System is buzzing with rocky material, ranging from microscopically small dust particles up to worlds the size of Texas. Huge lumps of ice, sprouting vast geysers of steam as they swing past the Sun, appear to us as comets. Far beyond the planets, swarms of icy and rocky chunks extend halfway to the nearest star.

All these kinds of cosmic rubble have one thing in common: they are debris from the formation of the Sun and planets.

# ASTEROIDS

Look at a map of the Solar System's planets, and you will see a yawning gap between Mars and Jupiter. Surely there should be a world here, reasoned the 18th-century German astronomer Johann Bode. He encouraged his colleagues across Europe to set up the "Celestial Police" to track down the missing planet.

In 1801, Giuseppe Piazzi, in Sicily, found an object, which he named Ceres. However, Ceres was so faint it had to be small. Over the next few years, the Celestial Police turned up more tiny worlds between Mars and Jupiter, and they were named asteroids ("starlike") after their appearance in a telescope. When astronomers started taking long-

*Goddess Ceres urges Piazzi to look through his telescope.*

exposure photographs of faint nebulae and galaxies, they often found the image spoilt by the trail of an asteroid that had strayed over the scene. As a result, asteroids were disparaged as "the vermin of the skies."

Astronomers have now discovered over 600,000 asteroids. In total, there could be as many as a billion space rocks circling the Sun between the orbits of Mars and Jupiter, in the region known as the asteroid belt.

Plenty are to be found outside the asteroid belt, as well. The Trojan asteroids share Jupiter's orbit, behind and ahead of the giant planet. Others have looping paths that bring them closer to the Sun, crossing the orbits of Mars and the Earth.

## ASTEROID NAMES

With automated telescopes scanning the sky every night, astronomers are discovering a thousand new asteroids every week! Most just have catalogue numbers, but the biggest asteroids are named.

*Giuseppppe Piazzi*

At first, they were called after ancient goddesses, such as Ceres, Pallas, Juno and Vesta, but the pantheon soon ran out, and astronomers turned to prominent scientists, politicians, cities and countries for names. But the system ran into abuse when the discoverers chose to honor their mistresses or even their cats. Now, the names are decided by the International Astronomical Union (so we feel very honored that we are "up there" as Asteroid 3795 Nigel and Asteroid 3922 Heather).

# ASTEROIDS IN CLOSE-UP

In 1991, the space probe Galileo took the first close-up image of an asteroid, as it sped past Gaspra on its way to a rendezvous with giant planet Jupiter. Gaspra turned out to be a potato-shaped lump of rock, just 18 km long, too small for gravity to pull it into a round ball.

Galileo then passed the rather larger Ida, which turned out to have a tiny moon in orbit; astronomers have now found over 150 asteroids that have their own moons.

Spacecraft have now visited 10 asteroids, all of which are rough and irregular in shape. The Near Earth Asteroid Rendezvous-Shoemaker (NEAR Shoemaker) spacecraft landed on asteroid Eros, while the Japanese probe Hayabusa picked up some dust from Itokawa and brought it back to Earth for analysis.

The Dawn mission has orbited Vesta, to study the heavily cratered surface of this large asteroid in unprecedented detail. Its images show gullies running downhill that may have been carved by water.

The biggest asteroids are generally made of solid rock; Vesta even boasts a

## DWARF PLANET CERES

In 2006, astronomers reclassified the largest asteroid, Ceres, as a "dwarf planet." This means its gravity is strong enough to pull Ceres into a round ball—unlike the smaller odd-shaped asteroids—but it is not a fully-fledged planet, which would have sufficient gravitational power to fling all the other asteroids out of its orbit and reign supreme in the space between Mars and Jupiter.

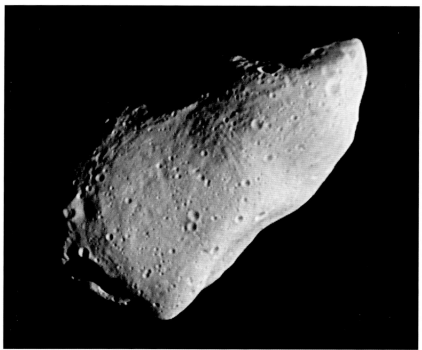

*Rocky, cratered asteroid Gaspra, imaged by the passing Galileo probe. It measures 18.9 x 10.5 x 8.9 km.*

metal core, like the Earth. About one-third the size of Vesta, asteroid Psyche is a lump of pure metal, undoubtedly the core of a big asteroid that was broken up in a collision long ago.

But most of the smaller asteroids are simply piles of primeval rubble, loosely held together by gravity. They are often dark in color, suggesting they are covered in rocks that are replete with organic molecules, the carbon-rich minerals that are the basis of life.

# OBSERVING ASTEROIDS

Asteroids are among the trickiest Solar System objects to observe. For a start, only one of them, Vesta, is bright enough to spot with the naked eye, and, even then, you will need a really clear, dark sky. However, several of them are within easy reach of a pair of binoculars.

To compound the difficulty, asteroids are constantly moving around. The best way to track them down is by using planetarium software on your computer, or an app for your phone or tablet. Astronomy magazines often map the positions of the brightest asteroids, month by month.

But it is the asteroids' motion that gives them away. As their name suggests, asteroids look like stars. Draw the pattern

## TOP 10 ASTEROIDS

| NAME | NAMED AFTER | YEAR DISCOVERED | DIAMETER/ LENGTH | COMPOSITION | NOTES |
|------|-------------|-----------------|------------------|-------------|-------|
| Ceres | Patron goddess of Sicily | 1801 | 975 km | Carbon-rich | Largest asteroid; also classed as "dwarf planet" |
| Pallas | Goddess of wisdom | 1802 | 582 km | Carbon-rich | Orbit highly tilted (34°) |
| Vesta | Goddess of the home | 1807 | 573 km | Stony | Huge crater at south pole |
| Psyche | Cupid's lover | 1852 | 240 km | Iron-nickel | Could satisfy Earth's metal use for millions of years |
| Kleopatra | Queen Cleopatra | 1880 | 217 km | Metal/stone | Shaped like a dog bone |
| Ida | Greek nymph | 1884 | 56 km | Stony | First asteroid found with a moon |
| Eros | God of love | 1898 | 34 km | Stony | First near-Earth asteroid |
| Hektor | Trojan war hero | 1907 | 93 km | Carbon-rich | Shares Jupiter's orbit |
| Phaethon | Son of Greek Sun-god | 1983 | 5 km | Carbon-rich | Passes closest to Sun; source of Geminid meteors |
| 2010TK$_7$ | Catalogue number | 2010 | 0.3 km | Not known | Shares Earth's orbit |

*At 525 km across, Vesta is the second-biggest asteroid, and also the brightest. This image was captured by the orbiting Dawn spacecraft.*

of stars where you suspect an asteroid is hiding, then look again the following night. The "star" that has moved is your prey.

## ASTEROID SCIENCE FROM THE BACKYARD

Amateur astronomers equipped to measure the brightness of celestial objects are making serious contributions to asteroid research. By following an asteroid's changing brightness, they can work out how long it takes to spin around. In the case of Antiope, the observations proved that it is two asteroids spinning together in a tight gravitational tango.

By watching a distant star disappear and reappear as an asteroid moves in front and measuring the duration of this *occultation*, backyard astronomers can also determine the size and shape of the asteroid.

# IMPACT EARTH

At 9:20 a.m. on February 15, 2013, citizens of the Russian city of Chelyabinsk thought that the end of the world had come. A brilliant fireball flashed across the sky and exploded above the heads of the horrified residents. It destroyed buildings and injured 1,500 people with flying glass.

Was it a nuclear missile? No, it was a 10,000-ton meteorite that had strayed from the asteroid belt. The Chelyabinsk blast was a reminder that asteroids can escape from their home, through collisions or being pulled by Jupiter's gravity.

Asteroids passing close to our planet are called "near-Earth objects" (NEOs). We now know of almost a thousand NEOs bigger than 1 km across— large enough to destroy a city—but, fortunately, none is currently on collision course with the Earth!

When we do find a rogue asteroid heading our way, space engineers have various plans for deflecting it. They could pull it gently off course with an unmanned spacecraft, push it away with brilliant laser light or, if time is pressing, blast it into a different orbit with a nuclear explosion.

*The Chelyabinsk fireball streaks across a clear sky*

*Landing spot for the Chelyabinsk meteorite — in a frozen lake*

## MASS EXTINCTIONS

Sometimes, a cosmic rock far bigger than the Chelyabinsk asteroid hits our planet — as the dinosaurs learned to their cost. Some 65 million years ago, an asteroid 10 km in diameter smashed into the Earth in what is now Mexico. It blasted out a crater 180 km across and threw up gigantic tsunamis along with shock waves and a blast of heat that seared the world. Large land-living animals were wiped out, including the dinosaurs.

Asteroids this size impact Earth every 100 million years or so. Previous hammer-blows from space may have caused the major mass extinctions in the past, when most species of life were wiped out. Next time it happens, the human species will be in the firing line.

# METEORITES

In 1938, a Nazi scientist traveled to Tibet, and brought back an ancient Buddha engraved with a swastika. When this metal statuette reappeared in 2007, scientists made an even more remarkable discovery: the statuette had been carved from a metal ingot that had fallen from space long ago. This Buddha was a true messenger from the heavens.

It was just one of thousands of meteorites found all over the Earth. Most meteorites are small chips off asteroids, and they provide a cross-section through their parent body. The main types are:

## STONES

Most meteorites are simply rocks smashed from the stony outer parts of an asteroid. Some stony meteorites resemble Earth's basalt: they were blasted out of Vesta by an ancient collision that left a giant crater on the asteroid. Others contain *chrondrules*, small drops of originally molten rock from the fiery birth of the Solar System.

## IRONS

People have always valued metal meteorites: the Inuit made tools and harpoons from the 31-ton Cape York meteorite. These lumps of an iron-nickel alloy are fragments from the metallic core of a completely disrupted asteroid.

## STONY-IRONS

As the name suggests, these are a mixture of stone and iron-nickel metal.

## CARBONACEOUS CHONDRITES

The ultimate prizes for scientists are "carbonaceous chondrites": carbon-rich rocks from an asteroid's surface that have remained unchanged for billions of years. They comprise a time machine to the very birth of our Solar System.

You can see meteorites in major museums or buy fragments for yourself. However, meteorites that are unusual or have an interesting history (like fragments of the Chelyabinsk meteorite) can fetch astronomical prices. Some are literally worth more than their weight in gold.

But, in many ways, every meteorite is priceless. It's the only object you can ever hold that's never been part of planet Earth.

*Arizona's Meteor Crater was blasted out by an iron meteorite 50,000 years ago. It measures 1.2 km across.*

# PLUTO AND THE KUIPER BELT

"Save Pluto!" cried the T-shirts, and "Stop Planetary Discrimination"; bumper stickers read "Honk if Pluto is still a planet!"

It was 2006, and passions were running high about a tiny world far out in space. American astronomer Clyde Tombaugh had discovered Pluto in 1930, and it was immediately hailed as "the ninth planet." However, in 2006, the International Astronomical Union demoted the little world. Now there are only eight planets, and the roll-call stops with Neptune.

What had changed? Astronomers had discovered that Pluto is not alone; keeping it company are thousands of small frozen worlds. It is like an icy version of the asteroid belt, beyond the large planets. Two astronomers, Kenneth Edgeworth and Gerard Kuiper, had predicted just such a belt: it is generally now called (a bit unfairly) the Kuiper Belt.

## PLANET OF DISCORD

In 2005, Mike Brown in California discovered a body later called Eris (meaning "discord"), which is slightly bigger than Pluto. Undoubtedly, the Kuiper Belt holds many more worlds like Eris. Were they all to be called planets, giving an ever-increasing total of perhaps 50 or more?

Astronomers voted instead to limit the word "planet" to only the biggest worlds of the Solar System. Pluto and Eris are now "dwarf planets," like the dwarf planet Ceres in the asteroid belt.

There are probably 100,000 Kuiper Belt objects bigger than 100 km in size, and billions altogether.

*Pluto (left) and its largest moon Charon. Pluto has five known moons.*

## PLUTO'S MOONS

For such a small world, Pluto has an amazingly large family. Charon was discovered on photographic plates in 1978. It is very large for a moon, half the size of Pluto itself, so astronomers refer to Pluto-Charon as a "double dwarf planet." Since then, astronomers using the Hubble Space Telescope have found four more tiny moons, bringing the total to five. They were all probably formed when another body whacked into Pluto, spraying out icy material that condensed to form the moons.

# COMETS

In Shakespeare's play *Julius Caesar*, the Roman leader's wife, Calpurnia, warns her illustrious husband that a blazing comet is foretelling his sticky end: "When beggars die, there are no comets seen; the heavens themselves blaze forth the death of princes."

Around the world, people have regarded comets as symbols of doom. The Chinese saw them as "broom stars," sweeping away the established order.

Appearing like a glowing curved sword suspended in the night sky, a comet is both beautiful and mysterious; many people

*In early March 1986, Halley's Comet returned to shine over the dramatic stone sculptures on Easter Island.*

## HALLEY'S COMET

When he began to analyze old comet observations, Halley found that a brilliant comet he had seen on his honeymoon in 1682 was following exactly the same orbit as the comets of 1531 and 1607.

Halley concluded this was the same comet, making repeated visits to the

*Halley's Comet as depicted on the Bayeux Tapestry, following its appearance in 1066*

Sun. Working forward, he calculated it would reappear in 1758. When the comet materialized that year, long after his death, it assured that his name would live on in the heavens.

Halley's Comet is a "once in a lifetime" experience, gracing our skies every 76 years. It was far from the Earth—and thus faint—on its last visit, in 1986. Expect a better display when Halley's Comet returns in 2061, and a really spectacular show in its 2134 appearance.

still remember the bright comet Hale-Bopp which hung in our skies for weeks in 1997.

Early scientists thought comets were emanations of gas in the atmosphere, until Danish astronomer Tycho Brahe proved the comet of 1577 lay beyond the Moon. The British scientist Isaac Newton insisted that comets, like planets, must obey his newly invented law of gravity; his friend Edmond Halley did the tedious calculations. Halley found that comets follow long, looping orbits, starting far beyond the planets of the Solar System, and potentially bringing them past the Sun again and again.

# A COMET'S TALE

A comet begins its life as a ball of ice and dust, orbiting far from the Sun in a giant swarm of cosmic icebergs that surrounds the Solar System, and stretches halfway to the nearest star. First predicted by the Estonian astronomer Ernst Öpik and Dutchman Jan Oort, the home of the comets is now called the Oort Cloud.

Occasionally, the gravity of a passing star, or the gravitational effects of the Milky Way galaxy, may dislodge a comet from the Oort Cloud, so it begins to fall towards the Sun.

As it gets closer to our Star, the Sun's heat makes the ice in this small, solid nucleus evaporate. They grow into a vast glowing "head" of gas and tiny specks of dust, called a coma. In 2007, the tenuous coma of Comet Holmes was as large as the Sun.

The solar wind—electrically charged particles perpetually streaming out from the Sun—buffets the gas in the comet's coma, and pushes it out into a long, straight fluorescent tail. At the same time, the dust particles are propelled outwards by the pressure of the Sun's light, to create a second, curved tail.

In color images of comets, you can see the gas tail glowing blue, while the dust tail is yellowish. They both point more or less away from the Sun; when a comet is heading back into space again, it is traveling tail-first.

If a comet has passed the Sun many times, like Halley's Comet, astronomers can predict quite accurately how bright it will be. However, this is much more difficult with a first-time comet. When Comet ISON was found in 2012, astronomers predicted a brilliant "Comet of the Century," but it brightened disappointingly slowly, and broke up altogether as it swept close to the Sun in 2013.

*Brilliant Comet Hale-Bopp blazes a trail across the skies of 1997. It has two tails: the bluish gas tail points directly away from the Sun, while the dust tail curves gently.*

# COMETS IN CLOSE-UP

In March 1986, a European space probe called Giotto took a death-defying plunge right through the coma of Halley's Comet at almost 100 times the speed of a rifle bullet. The comet's dust particles scoured the spacecraft, but it survived to fulfil its unique mission: to take the first close-up views of the solid nucleus at the comet's heart.

Giotto found a peanut-shaped ball of ice and rock, about 15 km long, spewing out geysers of steam and dust. To astronomers' surprise, this "dirty snowball" was almost black, with its surface coated in dark tarry compounds.

Since then, spacecraft have imaged five other comet nuclei. The Stardust mission also scooped up dust particles as it swept

## TOP 10 COMETS

| NAME | YEAR DISCOVERED | LAST PASSED SUN | ORBITAL PERIOD | NOTES |
|------|------|------|------|------|
| Halley | 240 BC | 1986 | 76 years | First orbit calculated |
| Great Comet | 1106 | 1106 | 1,000 years? | Brilliant sun-grazer: "parent" of 1882 and Ikeya-Seki comets |
| Lexell | 1770 | 1770 | 6 years | Closest approach to Earth (now lost) |
| Encke | 1786 | 2013 | 3.3 years | Shortest-period comet |
| Great Comet | 1882 | 1882 | 800 years | Brightest comet known (more brilliant than Full Moon) |
| Great January Comet | 1910 | 1910 | 57,000 years | Outshone Halley in 1910 |
| Ikeya-Seki | 1965 | 1965 | 1,000 years | Brightest comet of 20th century |
| Shoemaker-Levy 9 | 1993 | 1994 | 2 years (around Jupiter) | Crashed into Jupiter |
| Hale-Bopp | 1995 | 1997 | 2,500 years | Record naked-eye visibility (1.5 years) |
| ISON | 2012 | 2013 | New | Broke up as it grazed the Sun |

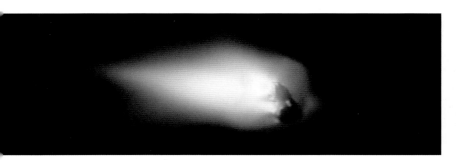

*Up close and personal to Halley's Comet: Giotto's view of the comet's dark nucleus*

through Comet Wild 2 and returned them to Earth for analysis. The Deep Impact mission probed Comet Tempel 1 in spectacular fashion, by smashing a heavy copper ball into the nucleus to see what lies inside.

In 2014, the Rosetta mission settled into orbit around Comet Churyumov-Gerisimenko, to observe the nucleus closely as it is warmed by the Sun's heat. The spacecraft discovered that the coment — just a few kilometers across — is shaped like a rubber duck. The "body" and "head" may be separate comet nuclei that have fused together. Rosetta has dropped a lander to explore the comet's surface.

# OBSERVING COMETS

A comet is among the most incredible sights you can see in the sky — especially a brilliant specimen that dominates the heavens, like Comet Hale-Bopp of 1997. The downside is that naked-eye comets are few and far between, and unexpected.

Keep an eye on the astronomical media and science websites to be forewarned of a potentially bright comet; they will also provide predictions and star charts. Most planetarium software and apps will automatically update with a new comet's positions, as well as showing the paths of old worn-out comets that have passed around the Sun many times and are now visible only in binoculars or a telescope.

To observe a comet, there's one golden rule: you must have a really dark sky. A comet's diffuse light is drowned out by even a hint of light pollution. You will get

## COMET DISCOVERY

A new comet can appear anywhere, at any time, and many have been found by amateur astronomers. Traditionally, a comet is named after its discoverers; the brilliant comet of 1997 was discovered independently by veteran comet-hunter Alan Hale and the rookie astronomer Tom Bopp.

Now, much of the romance is going out of comet hunting, as huge automated sky surveys are trawling comets from the depths of space. A lot of comets have an observatory name, like Comet PanSTARRS of 2013. Even so, Comet McNaught, one of the brightest comets of recent centuries, was discovered in 2006 by vigilant astronomer Robert McNaught while he was searching for near-Earth asteroids.

*An amateur observer with a small telescope studies Comet Hale-Bopp. In the northern hemisphere, it outshone every star in the sky except Sirius.*

the best view of the comet's elongated tail just by using your eyes. Binoculars are ideal for viewing its head, the coma.

Even the most powerful telescope will not reveal the tiny snowball of a nucleus, but you should see jets of gas erupting from the nucleus and pumping up the coma.

What is most fascinating is that you do not know what will happen next. The coma may suddenly puff up. You may see streamers in the comet's tail, or the solar wind might blow the gas tail away completely. As leading comet-discoverer David Levy says, "Comets are like cats; they have tails, and they do precisely what they want!"

# METEORS

"On the night of November 12-13, 1833, a tempest of falling stars broke over the Earth... At Boston, the frequency of meteors was estimated to be about half that of flakes of snow in an average snowstorm. Their numbers... were quite beyond counting." This is how astronomy writer Agnes Clerke described the stupendous celestial firework display that turned astronomers' attention to shooting stars.

## FIREBALLS

The most brilliant meteors, brighter than any of the stars and planets, are called *fireballs*. A fireball could be a solid lump of rock from an asteroid falling to Earth as a meteorite (pages 152-5) or a satellite (pages 176-7) burning up as it re-enters the atmosphere.

Until then, scientists had dismissed meteors as flashes in the atmosphere, like lightning. Now it was clear that meteors actually rained down from space. Today, we know that meteors are specks of dust from old comets. They run into the Earth's atmosphere, and burn up about 100 km above our heads.

Despite the brilliant show, these dust grains are surprisingly small, not much bigger than a granule of instant coffee. But, traveling at speeds of up to 250,000 kph, they pack a huge amount of energy that gets turned to light and heat as they burn up, leaving a long glowing trail of hot atmospheric gas behind them. Some also leave a luminous afterglow, a *train*, of dust.

Most meteors are a brilliant white, but some shooting stars display vivid colors. These hues may come from the heated atoms in the air, like the red and green colors of the aurorae, or from different minerals in the meteor itself; sodium gives a yellow tinge, while copper provides a

*A dazzling meteor streaks above the telescopes of the Atacama Large Millimeter/sub-millimeter Array (ALMA) in Chile's Atacama Desert.*

green hue. One of the authors, Heather, became an astronomer at an early age when she spotted a bright green shooting star.

# METEOR SHOWERS

O n August 10, 258 AD, a Christian deacon called Lawrence was martyred in Rome, and people in Italy still look to the skies with awe around that date, as the burning "Tears of St Lawrence" rain down from the sky.

These shooting stars are indeed burning, although they are solid grains of rock rather than teardrops. They are fragments from a comet called Swift-Tuttle, and we see them each year on the same date when Earth runs into Swift-Tuttle's trail of debris. As a result of perspective, the meteors seem to diverge from a specific

## METEOR STORMS

Occasionally, the Earth happens to hit a particularly dense stream of debris from a comet, and we witness a storm of meteors. In 1966, people in the western USA were astonished when the normally fairly sedate Leonid shower erupted into a veritable storm — with dozens of shooting stars *every second!*

## MAJOR METEOR SHOWERS

| NAME | PARENT BODY | DATE OF MAXIMUM | METEORS/HR | NOTES |
|------|-------------|-----------------|------------|-------|
| Quadrantids | Extinct comet 2003 EH1 | January 3-4 | 80 | Bright colorful meteors |
| Lyrids | Comet Thatcher | April 22 | 10 | High proportion of dust trains |
| Eta Aquarids | Halley's Comet | May 5-6 | 40 | Fast meteors |
| Perseids | Comet Swift-Tuttle | August 12-13 | 80 | High proportion of fireballs |
| Orionids | Halley's Comet | October 21-22 | 25 | Fast meteors; long-lasting trains |
| Leonids | Comet Tempel-Tuttle | November 17-18 | 30 | Very fast meteors |
| Geminids | Asteroid Phaethon | December 13-14 | 100 | Slow bright meteors |

*Raining meteors: this engraving of November 12, 1799, shows a storm of Leonid meteors over the ocean.*

point in the sky, known as the radiant: because this one lies in the constellation Perseus, the celestial firework display is called the Perseid meteor shower.

Each year, the Earth crosses the path of several different comets, including Halley's Comet, around these dates we are treated to showers of shooting stars. Like the Perseid shower, each is named after the constellation where its radiant lies. (One prolific meteor shower in January is named after a defunct constellation, Muralis Quadrans: this is now part of Boötes, lying near the tail of the Great Bear, Ursa Major.)

# OBSERVING METEORS

Backyard observers can make a highly valuable contribution to astronomy by monitoring meteor showers. Shooting stars can appear anywhere in the sky, at any moment, and the human eyeball—without a telescope—is the best instrument for detecting them.

To make the most of the celestial firework display, get well away from streetlights. Put yourself on a comfortable deckchair or sun-lounger, dressed warmly, and lie back to enjoy the action. Don't focus on the radiant position itself; you will see most meteors about 45 degrees away in any direction. The best time to look is after midnight, because you are then on the leading side of the Earth as it speeds around the Sun.

## Meteor party

Your eyes can only see part of the sky at any instant, so you will probably only catch half the hourly number listed on page 168. To snare as many meteors as possible and have more fun, organise a meteor party. Get together with some friends, and watch in different directions. One member of the party needs to record all the observations, using a red-light flashlight to see by, and an accurate clock to keep time.

When you see a meteor, shout out. You need to say how bright it appears, compared to nearby stars (see star magnitudes, page 204); and whether it is coming from the radiant of the shower, or is a sporadic meteor coming from a random direction. Ideally, hold a piece of string along the line of the meteor and memorize its location in the sky, then draw that path on a star map. Also, mention anything unusual, like a persistent glowing train.

Lastly, listen carefully, too. You may hear a crackling or swishing sound. It is probably because the meteor emits low-frequency radio waves that vibrate objects near you, like pine needles, dry grass or even the metal frames of your glasses.

*Leonid meteors, Spain, 2002. The "radiant" of the meteors is clearly visible — inside the "sickle" of Leo.*

# ZODIACAL LIGHT

On a crystal-clear evening, far from streetlights, you may spot a tall faint pyramid of light in the west after the Sun has gone down. It lasts a lot longer than the ordinary twilight glow and, instead of hugging the horizon, it forms a faint shining path stretching upwards. Look carefully, and you can see it runs through the constellations of the Zodiac (the line followed by the planets).

Consider yourself lucky. This is the elusive Zodiacal Light, and many astronomers have never been fortunate enough to view it. As well as needing a superbly clear and dark sky to reveal itself, it is only visible in the spring and autumn, when the Zodiac rises at a steep angle from the horizon.

You may also spot the Zodiacal Light in the east before dawn. The 12th-century poet and astronomer Omar Khayyám had an excellent view of this "false dawn," as he called it, over the Persian desert. It was worth celebrating with wine: "When false dawn streaks the east with cold, gray line, Pour in your cups the pure blood of the vine."

The Zodiacal Light is a fog of tiny particles that fills the Solar System in a flat pancake shape, out to the orbit of Mars.

## GEGENSCHEIN

The Zodiacal Light stretches, very faintly, all around the sky. At the point directly opposite the Sun, there is a large, slightly brighter patch called the Gegenschein (German for "counter-glow"), where sunlight is reflected back more intensely, like car headlights catching a highway sign. You will only see the Gegenschein when it is high in the sky and away from the much brighter Milky Way, during late autumn or early spring.

*The cone of the Zodiacal Light illuminates the desert skies of Chile's Cerro Paranal, home to Europe's Very Large Telescope.*

These dust particles come from old comets and asteroids that have collided. As our planet orbits the Sun, it scoops up around 40,000 tons of this space dust every year.

# NOCTILUCENT CLOUDS

During the summer months, you may be fortunate enough to see the most ghostly apparition in the night sky, noctilucent clouds. Their name is derived from the Latin for "night shining," and these icy clouds glow blue-white. They are illuminated by the Sun from below the horizon, and they're most commonly seen between latitudes of 50 and 70 degrees. Noctilucent clouds form above the Earth's poles, so you need look in that direction to see them.

So, why are we mentioning *clouds*, in a book on astronomy? First, these are the Earth's highest cloud layers. At an altitude of 80 km, they are right on the edge of space. Second, cosmic dust might be responsible for producing them.

At this frigid altitude, water vapor freezes into microscopic ice crystals, but it needs to condense onto solid nuclei. Many scientists believe that the nuclei in this case are particles of cosmic dust, drifting down into the Earth's atmosphere from space.

But the origin of noctilucent clouds is still controversial. They were first seen in 1885, soon after the eruption of the volcano Krakatoa in the eastern Indian

Ocean, so perhaps the nuclei are actually volcanic dust. Other scientists ascribe them to the Industrial Revolution, and the resulting increased pollution of the atmosphere. Another theory blames emissions from rockets blasting into space.

Whatever their cause, noctilucent clouds are becoming more common, so we can expect more of these glorious iridescent additions to the summer sky.

*A dazzling display of noctilucent clouds, photographed at midnight over Bragg Creek, Alberta, Canada.*

# SATELLITES

You will not be out observing for long before you spot a "star" that moves slowly and steadily across the sky, perhaps brightening and dimming, until it eventually fades from sight. What you are seeing is one of the thousands of satellites that humans have launched into space and are now orbiting Earth.

Most satellites are in "Low Earth Orbit," just a few hundred kilometers up. You see them when they catch the Sun's light, like very high-flying aircraft. When you are watching a satellite in the evening, it will usually travel into the Earth's shadow and fade away. Before sunrise, on the other hand, you can see a satellite suddenly *appear*, as it comes out of shadow and into the dawn sunlight.)

Generally, satellites travel in an orbit that takes them from west to east across the sky: that way, they gain extra speed from the Earth's rotation when they are launched. If you spot a satellite traveling north-south, it is one that has been put into an orbit where it can survey all of the Earth's surface to observe the weather,

## INTERNATIONAL SPACE STATION

If you see a brilliant light, brighter than any star, moving in a stately way across the sky, give it a wave. There'll be around six astronauts on board the International Space Station as it passes over you. The space station has been continuously occupied since November 2000, by space travelers of 15 different nationalities who are conducting research in weightless conditions, and also observing the Earth and the Universe from their unique perspective.

*The trail of the Echo communications satellite across the center of the Milky Way.*

monitor the environment or even to spy on other countries.

For predictions of the brighter satellites, you will need to check online for what can be seen from your own location. (See the websites on page 394.)

### IRIDIUM FLARES

Satellites change in brightness as the Sun's light glints off their irregular shapes. Most extreme are the Iridium communications satellites. Usually, they are too faint to be easily seen, but when the Sun catches one of their giant polished antenna panels, the satellites shine brilliantly for a few seconds—sometimes brighter than the planet Venus. These flashes, or "Iridium flares," are occasionally luminous enough to be visible in daylight.

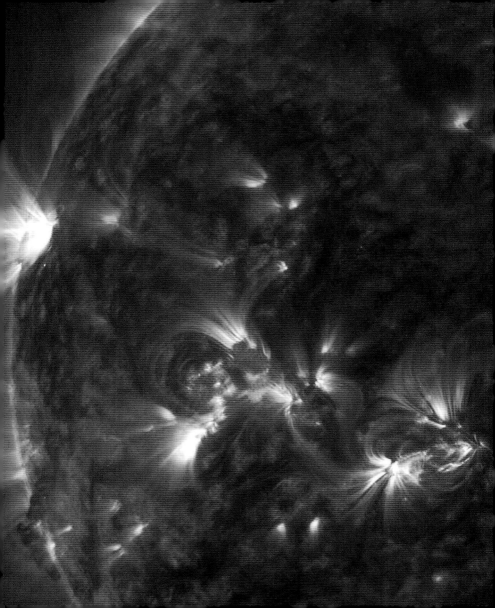

# CHAPTER 6
# SUN

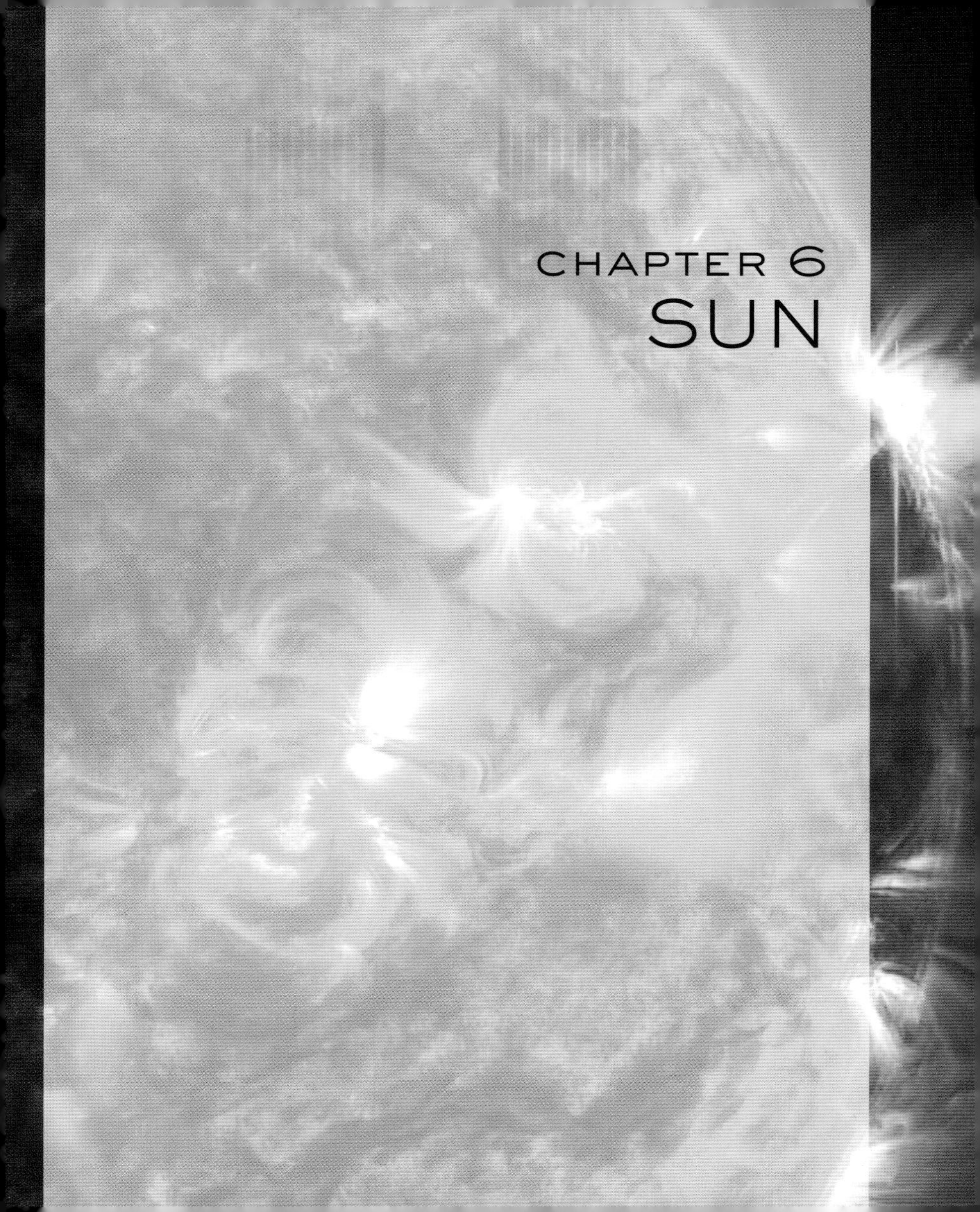

# INTRODUCTION

It is big and it is bright; and when it disappears from sight, we are plunged into the cold and confusion of night. It is no wonder that people throughout history and around the world have revered the Sun as a god.

A sculpture of a horse pulling the Sun chariot, found in a Danish bog, dates back almost 3,000 years. The Aztecs offered human sacrifices to appease Tonatiuh, their deity of the Sun. On a more cheerful note, an Australian Aboriginal tribe sees the Sun as a goddess, born on the dark Earth, who ascended to the sky with a burning torch.

Certainly, the Sun is the most important object in our part of the cosmos. As well as light, it provides us with the heat that prevents the Earth from freezing to the utter cold of space. The Sun's gravity keeps us in orbit at a safe distance and controls all the other planets, as well as the smaller bodies of the Solar System out to the most distant comets.

The superlatives about the Sun just go on and on. It is a thousand times more massive than all the planets put together. It is big enough to contain one million Earths. Its center is at a temperature of 27 million°F (15 million°C). In this inferno, nuclear reactions are creating vast amounts of energy, as hydrogen is converted to helium; the Sun is literally a hydrogen bomb running in slow motion.

However, if we step back from the Solar System, we find that the Sun is far from unique. It is a star, just like the thousands of others we see in the night sky. The Sun is middle-aged in astronomical terms, born some 4.6 billion years ago. It is also middle-of-the-road; many stars are hotter and brighter or dimmer and cooler.

## THE SUN: VITAL STATISTICS

| | |
|---|---|
| Diameter | 1.4 million km |
| Mass | 333,000 Earths |
| Rotation period | 25 days |
| Surface temperature | 9,900°F (5,500°C) |

*The life-giving Sun — our local star*

To us, though, the Sun is special.
It is the only star we can see up close
and without it, we would not be here.
Awesome.

# DAYTIME STAR

The Sun governs our lives. We wake in the morning around the time of sunrise, thanks to a bodily hormone that responds to its light. We feel drowsy, and fall asleep, when the Sun sets and its life-giving illumination disappears.

The ancient Egyptians even gave different names to the Sun at various times of the day. At sunrise, he was Khepri, a deified scarab beetle pushing his ball of dung, the Sun, above the horizon. During the day, he was glorious

*Celebration of the Sun: this wall-painting, in Egypt's Valley of the Kings, depicts Sun-god Ra and goddess Maat.*

*Passing the time of day: a sundial marks the hours as the Sun travels across the sky*

Ra, the king of the gods, traveling across the sky in his solar boat. The setting Sun was Atum, the creator of the Universe.

We now know the Sun itself is not moving; the Earth is turning round once every 24 hours and the Sun seems to move in the opposite direction in the sky, rising on the eastern horizon and setting towards the west.

Our 24-hour day has been inherited from the Egyptians. When the Sun was below the horizon, they believed it went through 12 "gates," (hours) of night. They divided the daytime up similarly, resulting in 24 hours as a single day.

The Egyptians also invented the sundial, which was an important way of measuring time right up till the 19th century. As the Sun appears to move across the sky during the day, it throws the shadow of a rod (the gnomon) onto the base of sundial, which is marked with hours and minutes.

Because the Sun crosses the sky at different heights during the year, the gnomon must be tilted at just the correct angle (equal to the dial's latitude) to indicate the time accurately at all seasons.

# ORBITS AND LEAP YEARS

It was perhaps the biggest shock in the whole of history: the solid Earth under our feet is actually careening through space faster than a rifle bullet. We are all familiar with the idea today, but the Polish canon Nicolaus Copernicus was challenging almost universally held beliefs when, in 1543, he argued that the Earth circles the Sun.

We complete one orbit around our local star every year. Inconveniently, the year is not an exact number of days: just over 365 days elapse before we lap the

## NEAR AND FAR

The Earth's path around the Sun is not a circle, but an egg-shaped ellipse. We are closest to the Sun (at perihelion) on around January 4 and farthest away (at aphelion) on about July 5.

Sun and return to our starting position. Julius Caesar realized that, if we simply used a calendar of 365 days, our dates would quickly get out of sync with the seasons. When visiting Cleopatra, he consulted the Egyptian astronomers, who proposed that every fourth year should be a "leap year" — when February would have an extra day.

This calendar assumes the year is exactly 365 days, 6 hours long; in fact, it is 11 minutes shorter. Over the centuries, the difference mounted up; the calendar again began to stray from the seasons. In around 1100, the Persian astronomer and poet, Omar Khayyám, devised a new rule for inserting leap years, and his calendar is still used in Iran.

In the West, Pope Gregory XIII refined the calendar in 1582 by decreeing that a century-year is only a leap year if it is divisible by 400 (rather than 4). Workers across Europe rioted

*Persian poet, astronomer and bon viveur Omar Khayyám, who — in around 1100 — developed a calendar comparable in accuracy to the one we know today. Khayyám's calendar is still used in Iran.*

when the new calendar was introduced, as it removed several days at a stroke, along with their pay, but the Gregorian calendar will stay in line with the seasons for at least 8,000 years to come.

# SEASONS

Winter is cold. Summer is hot. These seasonal changes have nothing to do with the Earth's distance from the Sun — they happen because our world is tilted over. In June, the northern hemisphere is tipped towards the Sun. People here see the Sun higher in the sky, and its heat falls more directly on the ground, giving the north the warmth of summer. At the same time, the southern hemisphere is tilted away, and countries there experience winter's cold and dark.

Six months later, the situation is reversed. Southern countries experience

*On Midwinter's Day, the Sun rises behind the leftmost tower of the 13 making up the newly-discovered solar observatory in Chankillo, Peru. Over the months, it rises behind each of the towers in turn, marking the progress of the year.*

summer as they bask in the heat and light of a Sun that is high in the sky while the northern hemisphere is cold.

As the seasons change, we see the direction of sunrise and sunset move along the horizon. Over 2,000 years ago, people living at Chankillo, in Peru, set up a series of towers on a hillside, where the position of sunset acts as a calendar to mark the passing months.

The Sun rises (and sets) at its farthest north on around June 21, which is the Summer Solstice in the northern hemisphere and the longest day of the year for that half of the world. Conversely, it rises and sets farthest south around December 22 (the northern Winter Solstice). For the southern hemisphere, these two dates are the Winter Solstice and Summer Solstice, respectively.

Roughly halfway between, the Sun lies over the Earth's Equator. It rises due east; and sets due west. These dates (around March 20 and September 22) are called the Equinoxes because day and night are of equal length.

## WINTER TEMPLES TO THE SUN

At Newgrange, in Ireland, the rising Sun on Midwinter's Day shines straight up a long passage deep into the ancient burial mound. As the Sun sets the same day, it illuminates the interior of the prehistoric burial site of Maes Howe, in Scotland's Orkney Islands. Many other stone structures around the world are also lined up with the Sun on the shortest day, when ancient people would pray for the return of its warmth and light. Modern-day Druids gather at Stonehenge in southern England to celebrate the Summer Solstice, but they are probably six months out. It is most likely this great monument, too, was aligned with sunset at Midwinter.

# OBSERVING THE SUN

The Sun is the easiest astronomical object you can observe. You do not have to stay up late, brave the cold of night or worry about light pollution. But, unlike any object in the night sky, the Sun is dangerous. Its blindingly bright light is matched by searing amounts of heat.

*Safely projecting the Sun's image onto a piece of card*

### ECLIPSE GLASSES

When there is an eclipse around, shops and astronomical societies sell or dispense special cardboard sunglasses with very dark lenses, for observing the partial phases. You can use these "eclipse glasses" at any time to view the Sun directly with the naked eye.

### PROJECTION, USING BINOCULARS OR A TELESCOPE

Point the instrument towards the Sun, without looking through it, by minimizing the size of the shadow it is throwing on the ground. Put a piece of

It's dangerous to stare straight at the Sun. NEVER observe the Sun directly with binoculars or a telescope, or you may blind yourself permanently. On these pages you will learn about some safe ways to observe our local star.

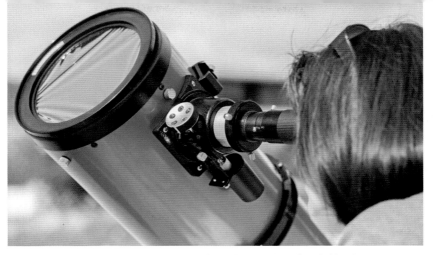

*Observing the Sun in safety, with an aluminium-coated Mylar filter placed over the end of the telescope*

white card behind the instrument, as a screen, and focus the instrument until the image of the Sun is sharp.

### SOLAR FILTER

These come in two kinds. Never use a filter at the eyepiece end of a telescope: the Sun's concentrated heat may crack it, and seriously damage your eye. The only safe solar filter is a special type of aluminium-coated plastic called Mylar (obtainable from telescope suppliers), which you fit over the *front* end of the telescope. Ensure there are no scratches or holes, and fix the filter firmly so

it cannot fall off. You can now look through the telescope.

### SOLAR BINOCULARS AND TELESCOPES

Special binoculars with a built-in dark filter allow you to observe the Sun directly (but nothing else, as they dim the view to the point of invisibility.)

You can buy a solar telescope with a narrow-band filter that is "tuned" to the light from one particular kind of atom, usually hydrogen. This will reveal astounding detail of the Sun's lower atmosphere.

# INSIDE THE SUN

Our local star is a hot ball of gas throughout. Although the Sun has no solid surface, we cannot see inside because this searing gas is opaque. However, astronomers have worked out the Sun's internal structure by analyzing its energy output, and from the way its surface vibrates, just as geologists can probe the Earth's interior by studying earthquakes.

They have found that the Sun has several distinct layers:

## CORE

This is a place from Hell. The gas in the Sun's center is so compressed that it's 10 times more dense than gold. The temperature is so high that hydrogen is fused into helium, in nuclear reactions that convert four million tons of matter into energy every second. Scientists on Earth can monitor these reactions directly, by detecting subatomic particles called neutrinos.

## RADIATIVE ZONE

Energy from the core travels outwards in the form of high-energy radiation, mainly gamma rays.

## TACHOCLINE

Swirling gas in this thin layer generates the Sun's magnetic field.

| | DISTANCE FROM CENTER | DENSITY* | TEMPERATURE |
|---|---|---|---|
| Core | 0–175,000 km | 150 | 27 million°F (15 million°C) |
| Radiative zone | 175,000–500,000 km | 20–0.2 | 12.6–3.6 million°F (7–2 million°C) |
| Tachocline | 500,000 km | 0.2 | 3.6 million°F (2 million°C) |
| Convective zone | 500,000–693,000 km | 0.2–0.000,000,2 | 3.6 million–9,900°F (2 million–5,500°C) |
| Photosphere | 693,000 km | 0.000,000,2 | 9,900°F (5,500°C) |

* compared to water

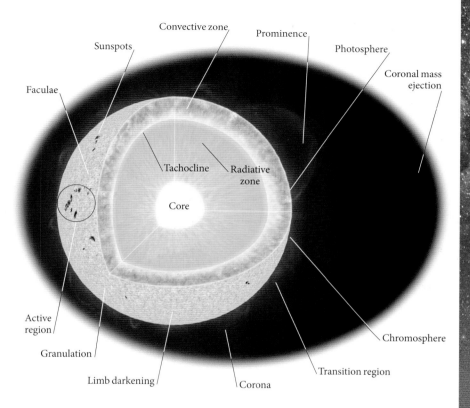

Convective zone

Sunspots

Prominence

Photosphere

Coronal mass ejection

Faculae

Tachocline

Radiative zone

Core

Active region

Chromosphere

Granulation

Transition region

Limb darkening

Corona

## CONVECTIVE ZONE

In the outermost region of the Sun, the gas is moving up and down in great currents, like water boiling in a pot.

## PHOTOSPHERE

At the top of the convective zone, the hot gas becomes transparent; this is the "surface" that we see.

# SOLAR SURFACE

The ancient Chinese were observing dark markings on the Sun over 2,000 years ago, even if they interpreted sunspots in terms of astrology: "A crow appears within the Sun: governance is chaotic; great drought."

In fact, there is a lot more to the apparently bland face of the Sun, the photosphere, than meets the eye.

### LIMB DARKENING

Even a pair of eclipse glasses shows that the Sun appears brightest at the center. The glowing photosphere is not a solid

Before observing the Sun, first read the warnings and guidance on pages 188–9.

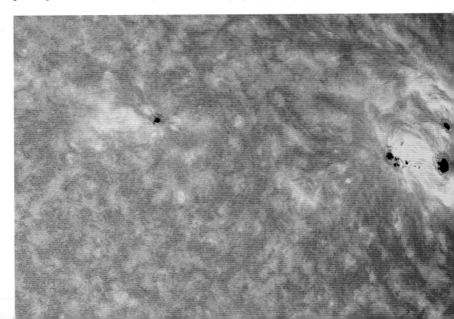

surface; towards its edge, we look less deeply into the Sun, and observe its cooler and dimmer upper regions. Under higher magnification, you may spot brighter patches near the edge, called faculae.

## GRANULATION

Close-up, the Sun's surface resembles a bowl of rice. Each "grain" is the top

*Sunspots pockmark the face of the active Sun. Although they appear darker than the photosphere — which is at a temperature of 9,900°F (5,500°C) — sunspots can reach a respectable 7,200°F (4,000°C).*

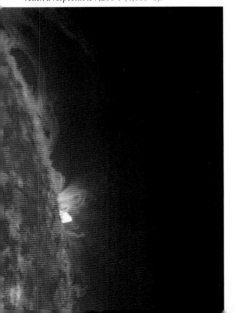

of a stream of hot gas, welling up from within. Time-lapse observations show the granulation is in constant motion, like a pot of boiling oatmeal.

## SUNSPOTS

A sunspot is a small dark blemish on the Sun's beaming face, although, on this huge globe, everything is relative. Many sunspots are as large as the Earth. Each has a dark center, the umbra, fringed by the less intense penumbra. Sunspots often appear as a pair or in a larger group, called an active region.

Here, the Sun's magnetism is breaking through the surface. It puts a lid on the energy trying to bubble up from underneath, creating a cooler, darker region. However, a sunspot is far from completely black, it simply looks that way in contrast with the brilliant photosphere. If you could view a sunspot in isolation, it would shine as brightly as the Moon.

You can track the Sun's rotation by following a prominent sunspot from day to day.

# SUN'S ATMOSPHERE

Above the Sun's surface, the photosphere, it has a tenuous atmosphere that reaches way into space. In fact, the Earth orbits the Sun *within* its atmosphere. In 2012, the speeding Voyager 1 spacecraft finally exited the Sun's far-flung gases, at the heliopause, four times farther out than the orbit of most distant planet Neptune, and entered interstellar space.

## CHROMOSPHERE

The thin gases in the lowest layer of the atmosphere shine pink with the light from hydrogen atoms. The chromosphere is normally invisible, but you can view it with a special solar telescope (see page 188) or during a total eclipse.

You may see long streamers of this gas hanging above the chromosphere either as dark silhouettes or as glowing pink prominences at the Sun's edge.

## TRANSITION REGION

Rather like an induction hob in the kitchen, the transition region converts magnetic energy from the Sun's surface below into heat, raising the temperature of the corona, the outermost layer of the atmosphere, to 1.8 million°F (1 million°C).

## CORONA

The corona is invisible to ordinary telescopes (except during an eclipse), but its glowing magnetic loops form a

| | HEIGHT ABOVE SURFACE | TEMPERATURE |
|---|---|---|
| Photosphere | 0 km | 9,900°F (5,500°C) |
| Chromosphere | 0–2,300 km | 7,200–36,000°F (4,000–20,000°C) |
| Transition region | 2,300–3,000 km | 7,200–1.8 million°F (20,000 – 1 million°C) |
| Corona/ solar wind | 3,000–18 billion km | 1.8 million–180,000°F (1 million – 100,000°C) |
| Heliopause | 18 billion km | 180,000°F (100,000°C) |

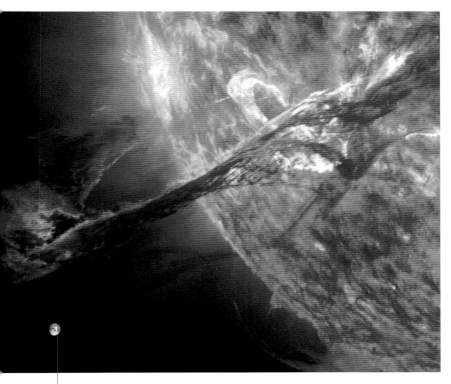

EARTH TO SCALE

*August 31 2012, a long filament of gas floating in the corona is jettisoned into space. This coronal mass ejection (CME) traveled at a speed of nearly 1,500 km/s towards Earth, triggering an aurora.*

stunning sight when imaged by satellites, like the Solar Dynamics Observatory, that observe the ultraviolet and X-ray emissions from the Sun.

The corona is so hot that its gases are constantly boiling away into space. Every second, the solar wind blasts away a million tons of the Sun's matter.

# SOLAR CYCLE

In the early 19th century, a German amateur astronomer, Heinrich Schwabe, carefully watched dark spots on the Sun every day for 17 years, thinking that he might see an unknown planet pass across its face. Instead, he discovered that the number of sunspots rises, and then falls again, over a period of 11 years.

The sunspot cycle is caused by changes in the Sun's magnetic field. Generated inside the Sun, at the tachocline, it threads upwards through the outer convective zone (see page 190–91). But the Sun's equator spins around faster than the gas at its poles, stretching out the magnetic field, like a rubber band, until something has to give.

## THE DANGER OF SOLAR STORMS

When a solar storm hit the Earth in 1989, it sent vast currents through the Canadian electricity grid, cutting off power to six million people in wintry Québec. Solar weather can also disrupt satnav systems and mobile phone networks.

A solar storm in January 1994 crippled two communications satellites. The Sun's violent weather also makes the Earth's atmosphere expand, so low-orbiting satellites might burn up. To date, solar storms have caused $4 billion-worth of damage to satellites.

Astronauts flying above the Earth's magnetic cocoon are in serious danger. A potentially lethal eruption on the Sun in 1972 fortunately fell between two Apollo missions to the Moon. But future Mars-bound astronauts could be in mortal danger from solar eruptions.

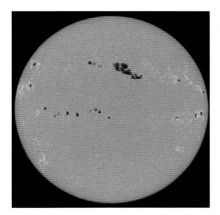

*At sunspot maximum, the Sun is dotted with multiple spots.*

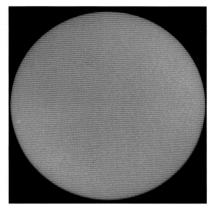

*The Sun's bland face at sunspot minimum.*

## MAXIMUM AND MINIMUM

Like a writhing piece of elastic, the magnetism breaks out through the surface, in a rash of sunspots. The first spots of a new cycle erupt near the Sun's poles; over the next few years, they appear nearer to the equator as our local star builds up to "solar maximum," when we may see a hundred sunspots simultaneously.

Eventually, the magnetic field and the sunspots die away again. At solar minimum, weeks may go past without any spots appearing on the Sun's face.

As well as having sunspots, the Sun undergoes a frenzy of magnetic activity. Where loops of magnetism touch, they explode, like two electric wires short-circuiting, in a brilliant flare. Magnetic eruptions high in the Sun's corona are not so obvious, but they can spew out even more dangerous clouds of superheated gas, called coronal mass ejections. Both kinds of solar storm eject streams of dangerous particles and radiation that wreak havoc through the Solar System.

# ECLIPSES OF THE SUN

A total solar eclipse is the most awesome of all sky sights. The brilliant Sun is eaten away, to be replaced by something you have never seen even in your wildest dreams: a grotesque dragon mask with a gaping black mouth, perhaps, or a gorgeous celestial flower.

Scientifically, we are seeing the Moon moving right in front of the Sun and blocking out its light. Purely by chance, the Sun and the Moon look the same size in the sky (the Sun is 400 times bigger, but it is also 400 times as far away) and the Moon can cover the Sun's glowing disc exactly in a *total* solar eclipse.

When the Moon is at its farthest from Earth, and appears smallest, we see a ring of the Sun's surface around the black silhouette, in an *annular* eclipse. More often, we witness the Moon covering only a portion of the Sun's surface, in a *partial* eclipse.

## OBSERVING A SOLAR ECLIPSE

You need to be within a very narrow band on the Earth to see a given total eclipse (see page 388–9 for forthcoming eclipses). Several tour companies organize eclipse trips, selecting a site where the weather should be good.

During the partial phase, use eclipse goggles (supplied by your tour operator) to protect your eyes from the Sun's heat. As soon as totality starts, and the Sun's brilliant surface is hidden, you can look at the Sun directly; then it is safe to use binoculars or a telescope, too.

The thin band of pink light around the Moon's silhouette is the chromosphere, with perhaps some prominences rearing upwards. Above stretches the Sun's faintly glowing outer atmosphere, the corona.

Look around to spot planets and bright stars in the darkened sky and then back to the glorious sight of

*Total eclipse of the Sun on October 24, 1995. The corona fans out beautifully against the dark sky, while the emerging Sun creates a "diamond ring" effect behind the Moon's limb.*

the Sun. The total eclipse ends with the Diamond Ring, the first speck of the brilliant photosphere appearing as a brilliant jewel in the ring of the chromosphere. Immediately put away binoculars and telescopes, and put on your eclipse goggles to watch our daytime star reappear.

# AURORAE

Scottish people have long been awed by the "Merrie Dancers": swirling red and green curtains and arches of light shimmying across the night sky.

According to folklore, they are due to sunlight reflecting off the polar cap.

In fact, the Sun is the cause, but in a completely different way. When its

magnetic field builds up to a crescendo of dark sunspots and giant loops of hot gas, every 11 years, it can hurl an avalanche of hot gas and radiation at the Earth (see page 197).

Fortunately, our planet's magnetic field acts like an invisible umbrella, protecting us from the worst of the solar storms, but electrical particles from the Sun can stream downwards at the Earth's magnetic poles. Impacting on the atmosphere, they light up gas atoms, oxygen in green and nitrogen in red, and we witness a spectacular display of the aurorae.

When the Sun is really stormy, the aurorae can be seen in regions as near the Equator as the Mediterranean and the southern parts of North America. However, this natural light show is usually seen only from regions nearer the poles: in the Arctic as the aurora borealis (Northern Lights) and near Antarctica, as the aurora australis (Southern Lights).

You can take special tours by plane or cruise ship to view this unique light show in its full glory, or stay inside a "glass igloo," where you can lie in the warmth of your bed and see the magnificent display unfolding over your head.

*In March 2008, the crew of the Endeavour Space Shuttle captured this sensational view of the aurora hovering above the Earth, on a mission to dock with the International Space Station.*

# INTRODUCTION

Apart from being an awesome sight, the vista of a transparent, starry sky has guided humankind throughout the millennia. As the Earth spins, the stars circle overhead, giving us a celestial clock. During the year, different constellations signal the seasons, and these tiny points of light in the sky have aided navigation since ancient times.

The first person to rank the stars according to their brightness was the Greek astronomer and mathematician Hipparchus. Born in around 190 BC, he spent his life making revolutionary discoveries in astronomy and trigonometry. Hipparchus compiled a star catalogue that was still in use over 1,500 years later. He characterized the stars into six brightness classes: one for the brightest, six for the faintest.

In 1856, the British astronomer Norman Pogson rationalized Hipparchus' classes into our modern "magnitude" system. On Pogson's scale, a first-magnitude star is 100 times brighter than one of the sixth magnitude. The (confusing) thing is that there are actually some stars brighter than the first magnitude — and these have negative magnitudes (like brilliant Sirius).

With a telescope, you can see even fainter stars. A typical amateur telescope, with a mirror 150 mm across, can show you stars as faint as magnitude 13. The orbiting Hubble Space Telescope, soaring above our obscuring atmosphere, picks out stars and galaxies of the 31st magnitude!

Light pollution, however, ruins our view of the heavens. In a city, you would be pushed to see stars fainter than magnitude 3, but on a brilliant night in the countryside, for instance at a designated Dark-Sky Park, you can see 3,000 stars down to magnitude six (the limit for the unaided eye).

*For a list of the brightest stars, see page 374.*

*The iconic constellation of Orion, the mighty hunter. Below his belt of three stars hangs his sword — bejewelled by the glowing Orion Nebula, a birthplace of stars.*

# NAMING THE STARS

Why do the brightest stars have such strange names? The reason is that the names date from antiquity, and have passed through down generations ever since. The original western star names, like the first constellations, were probably Babylonian or Chaldean, but few of these survive. The Greeks took up the baton after that, and the name of the brightest

star, Sirius, is one direct result. It means "the scorcher," because this star is nearest the Sun during the hottest days of the summer.

But Persian astronomers were largely responsible for the star names we have inherited today. Working in the so-called "Dark Ages" between the 6th and 10th centuries AD, they took over the naming of the sky, which is why so many stars' names begin with "al" (Arabic for "the"). Alioth, the first star in the "tail" of the Great Bear (Ursa Major), means "the fat tail of the eastern sheep," presumably because sheep were more common than bears in the deserts of the Middle East. Deneb, in Cygnus, also has Arabic roots; it means the tail (of the flying bird).

The most famous star in the sky has to be Betelgeuse, known to generations of schoolchildren as "Beetlejuice." It was gloriously interpreted to mean "the armpit of the sacred one." But the "B" in Betelgeuse turned out to be a mis-

## THE GREEK ALPHABET

| SYMBOL | NAME | SYMBOL | NAME |
|--------|---------|--------|---------|
| α | alpha | ν | nu |
| β | beta | ξ | xi |
| γ | gamma | ο | omicron |
| δ | delta | π | pi |
| ε | epsilon | ρ | rho |
| ζ | zeta | σ | sigma |
| η | eta | τ | tau |
| θ | theta | υ | upsilon |
| ι | iota | φ | phi |
| κ | kappa | χ | chi |
| λ | lambda | ψ | psi |
| μ | mu | ω | omega |

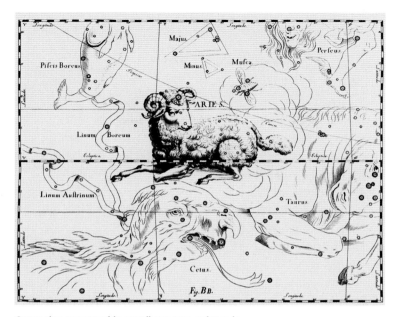

*Copper-plate engraving of the constellation Aries, as depicted in Johann Bayer's ambitious star atlas* Uranometria *in 1603.*

transliteration — and so we are none the wiser as to how our distant ancestors really identified this fiery red star.

Fainter stars in a constellation were seldom graced with names. Then, in 1603, the German astronomer Johann Bayer came up with the idea of naming stars after letters in the Greek alphabet, alpha for the brightest, beta for the second, and so on, followed by the constellation name in its genitive form (see pages 380-83). For a guide to the Greek alphabet, see the table on the previous page.

With the construction of ever-increasing sizes of telescopes, and the resulting discovery of more and more stars, astronomers had to resort to numbers. In all, over 300,000 stars were given HD numbers, in the honor of American astronomer and physician Henry Draper, whose widow funded the compilation of this giant catalogue.

# HOW FAR AWAY ARE THE STARS?

It is a big question. But astronomers knew that they had to tackle it if they were to learn anything about the nature of the stars themselves. Although they realized that measuring distances to the stars would be difficult, and would need precision telescopes, the idea behind the investigation was simplicity itself.

Look at a nearby star when Earth is in, say, its "June" position around the Sun, then observe it again in December. The star appears to "jump" against the background of more distant stars — an effect called "parallax." By measuring the size of the shift, and knowing the diameter of Earth's orbit, it is just a matter of trigonometry to work out the star's distance.

And that's exactly what German astronomer Friedrich Bessel did in 1838. He homed in on a faint pair of stars called 61 Cygni. He knew that the duo must be nearby because of their rapid motion across the sky. Over several years, Bessel scrutinized the pair. He succeeded in measuring the tiny parallax shift, and discovered that the star system lay a staggering 10 million milliom km away.

Unknown to Bessel, the first stellar distance had *already* been measured, but it was not published until 1839. Thomas Henderson, a Scottish lawyer-turned-astronomer, was observing half a world away, in South Africa. He measured the parallax of one of the most prominent stars in the southern hemisphere, Alpha Centauri, and calculated its distance as 40 million million km.

Meanwhile, Friedrich Struve, at the Pulkova Observatory in Russia, homed in on a beacon in the northern hemisphere, the star Vega. It turned out to be even farther away than 61 Cygni at 240 million million km away.

If it is any consolation, astronomers are no more able to comprehend these enormous numbers than anyone else. So they worked out a shorthand way to express them. Imagine light beams, speeding towards you from a star. Light travels at the speed limit of the Universe: 300,000 km per second. Astronomers measure the distances to the stars in terms of the time it takes their light to reach us.

For instance, the Sun's light takes 8.3 minutes to reach us, so our local star is 8.3 light minutes away. Alpha Centauri is 4.3 light *years* away, 61 Cygni is 11 light years distant and Vega lies 26 light years from our Solar System.

However, Vega is one of our *nearest* stellar neighbors. As time has gone on, astronomers have measured the distances to ever more remote stars, culminating in the work of the European spacecraft Gaia. Launched in 2013, it is measuring the position of stars up to a thousand times farther away than Vega.

*For a list of the nearest stars, see page 376–7.*

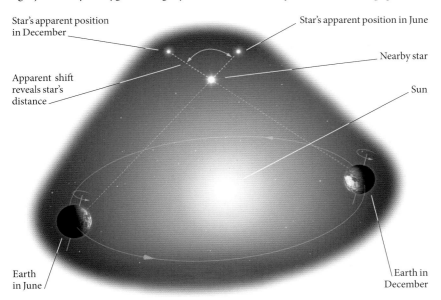

Star's apparent position in December

Star's apparent position in June

Nearby star

Apparent shift reveals star's distance

Sun

Earth in June

Earth in December

# WHAT IS A STAR?

Stars are the nuclear furnaces of the cosmos. Like the Sun, our very own star, they are powered by atomic fusion in their superheated cores. That is what makes stars shine. The energy floods out, creating a violent, active celestial creature. However, these fierce beasts give home to planets, now numbered in their thousands, some of which could bear life.

A star's story is a life story. Like us, they are born, they live and they die.

Stars are created in dark clouds of star-soot and gas that float around our Galaxy. Under the force of gravity, a cloud begins to curdle. Knots of gas collapse catastrophically. In the heat and fury of the nuclear reactions, fledgling stars are born.

Young stars light up their birthplace as a stunning nebula (the Orion Nebula is one of the most spectacular) with their powerful radiation searing their surroundings.

Stars range colossally in size, mass, luminosity and temperature. Our Sun is a middle-aged, middle-class, very average star. However, there are extremes out there. Spica, the brightest star in the constellation of Virgo, is 12,100 times more luminous than the Sun.

Indeed, the biggest stars are over a thousand times larger than the Sun. When they get to this stage, at the elderly red giant phase of their lives, these stars flop around unpredictably. A star that has become a red giant is running out of fuel: death is nigh.

A lightweight star will bow out gracefully, gently puffing off its atmosphere, leaving its core exposed as a dying white dwarf star.

Not so the heavyweights. Massive stars explode as supernovae, seeding the Universe with the elements of creation. A giant star is a phoenix; from its ashes arise the seeds that form a new generation of stars, planets and life.

*Close-up of our local star, the Sun. Loops of magnetised gas erupt from its stormy surface.*

# THE MESSAGE OF STARLIGHT

Scrutinize the stars on a dark, transparent night. You will discover that not all of them are plain, boring white. Take Orion's Betelgeuse, for instance, or Antares in Scorpius. Both of these stars are distinctly red — a sure sign that they are cool stars. Their surface temperature is around 5,400°F (3,000°C), as compared to 9,900°F (5,500°C) for our yellow Sun.

Contrast these stars with Rigel, another Orion resident. It shines a steely blue-white. This raging star is 125,000 times brighter than the Sun, and boasts a surface temperature of 21,600°F (12,000°C).

Star colors work like a stellar thermometer. The hottest stars are blue-white. White stars come next, then yellow, orange and red.

But what about the composition of the

## INSIDER'S GUIDE TO THE TOP 10 STARS

| NAME | DISTANCE | LUMINOSITY COMPARED TO SUN | TEMPERATURE | SIZE COMPARED TO SUN |
|---|---|---|---|---|
| Antares | 550 light years | 57,500 | 6,100°F (3,400°C) | 880 |
| Betelgeuse | 640 light years | 120,000 | 5,650–6,550°F (3,140-3,640°C) | 1,000 |
| Canopus | 310 light years | 15,000 | 13,200°F (7,350°C) | 70 |
| Deneb | 2,000 light years | 100,000 | 15,300°F (8,500°C) | 150 |
| Polaris | 325 light years | 2,500 | 10,800°F (6,000°C) | 46 |
| Proxima Centauri | 4.24 light years | 0.0017 | 5,400°F (3,000°C) | 0.14 |
| Rigel | 860 light years | 125,000 | 21,600°F (12,000°C) | 74 |
| Sirius | 8.58 light years | 25 | 18,000°F (10,000C) | 1.7 |
| Spica | 260 light years | 12,100 | 40,300°F (22,400°C) | 3.6 |
| Vega | 25 light years | 40 | 17,300°F (9,600°C) | 2.4 |

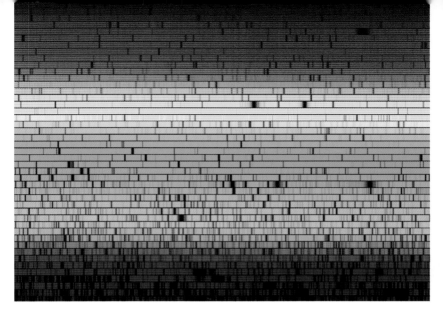

*Split into a spectrum of colors, and arranged in narrow horizontal bands, the Sun's light is besmirched by dark "lines" — the fingerprint of different elements.*

stars? In the early 19th century, a young Bavarian glassmaker named Joseph von Fraunhofer perfected the art of making exquisite glass. He made prisms, which split light into all the colors of a rainbow.

Applying his prism to the Sun's light, he discovered that the rainbow of colors was crossed by 574 dark, vertical lines. He found similar dark lines in the light from some of the brightest stars.

Half a century later in Germany, the scientists Robert Bunsen and Gustav Kirchoff were conducting an experiment on gases with a prism at its heart. Their "spectroscope," now a major tool in astronomy, revealed the meaning of Fraunhofer's lines: they were a signature of the chemical make-up of a star. Different elements absorb light from a star at different wavelengths.

Spectroscopy reveals the nature of the stars: a guide as to how they are born, live and die.

# DARK CLOUDS AND STAR BIRTH

The Aboriginal people of the Australian outback were blessed with skies so clear they lost their way among the multitude of stars. Instead of patterns of bright stars, they made constellations out of the *dark* patches in the sky. The most important is the Coalsack, a name given to it by European settlers, which the natives of Australia see as an emu, lying in wait for a possum perched in a tree (the Southern Cross, Crux).

Amazing as it may seem, the darkest patches in the sky, like the Coalsack or the Horsehead Nebula in Orion, are the spawning ground for brilliant stars.

The saga begins deep in space. It is not totally empty, anywhere in our Galaxy. Space is filled with tenuous gas, laced with "dust": microscopic specks of rock. The dust grains block out light from stars behind, creating what look like great empty voids in the Milky Way, most spectacularly in the constellation Cygnus, where the Milky Way is split in two by a streamer of dense dust.

Over time, gravity pulls this flimsy material together, into colossal dense clouds that appear even darker, like the Coalsack. But, deep inside, something is stirring. Denser knots of dirty gas coalesce, like sour milk curdling. Each of these knots shrinks under its own gravity, pulling it smaller and smaller. And as it shrinks, it grows steadily hotter. Astronomers call this dense ball of gas a protostar.

And then something miraculous happens. When the temperature in the center of the protostar soars to 18 million°F (ten million°C), hydrogen atoms knock together so fiercely that they combine to make helium. A nuclear furnace switches on. Energy surges out through the protostar, preventing it from collapsing.

Hidden deep inside its dark nursery, a star is born.

*Cosmic chess-piece: the Horsehead Nebula in Orion is a dark cloud poised to give birth to thousands of stars.*

# NEBULAE

When dark clouds collapse to form stars, it is as if the Christmas lights have been switched on. The once-black, glowering cloud of gas is turned into a brilliant, shining nebula. With their softly glowing light and lacy gaseous tendrils, lit from within by their newly born stars and fringed by delicate filaments of gas, nebulae are among the most beautiful sights in the cosmos. With their growing families of young stars, they are also the future of the cosmos.

### ORION NEBULA

The most iconic nebula in the sky has to be the Orion Nebula. Near "Orion's Belt" of three stars lies a fuzzy patch, which is easily visible to the unaided eye in dark skies. Through binoculars, or a small telescope, the patch looks like a small cloud in space.

It *is* a cloud, but at 24 light years across, it is hardly petite. Only the distance of the Orion Nebula, 1,300 light years, makes it seem so. Yet it is the nearest region to Earth where heavyweight stars are being born. This "star factory" contains at least 700 fledgling stars, which have condensed out of dark dust and gas clouds.

The most prominent are the four stars making up "The Trapezium" cluster. These are the red carpet "stars" of the Orion Nebula and are easily visible through a small telescope. They were born around 300,000 years ago and are babies when compared to the Sun.

These fractious infants stir up their environment. They, and their young companions in the nebula, kick up a stink. In their youthful restlessness, they spew out jets and punch their natal cloud with fierce stellar winds that shape the nebula.

The Orion Nebula represents only a fraction of the stellar activity going on in the Orion region. It is just the most prominent feature of the Orion Molecular Cloud, a constellation-wide

*Hubble Space Telescope close-up view of the Orion Nebula. Violent young stars are churning up their natal gas with powerful radiation and stellar winds, making it glow myriad colors.*

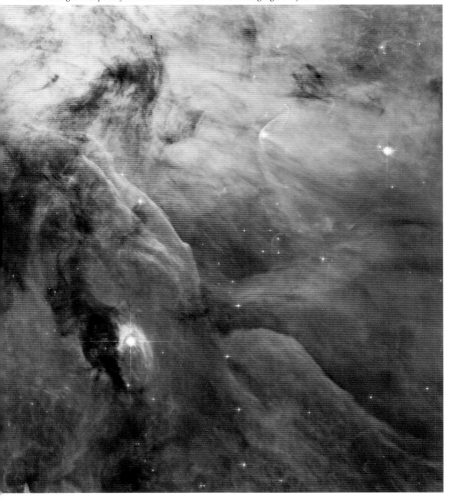

complex that includes the Horsehead Nebula (a dark cloud) and other nebulae.

The Orion region is a hotbed of star formation — and, with its generous supplies of interstellar gas, it is poised to supply our galaxy with stars and planets for the foreseeable future.

## CARINA NEBULA

Nebulae are the most glamorous A-list stars in the Universe. It is hard to single out one nebula from all the others, but the Carina Nebula, in the constellation of the same name, is exceptional. Some 9,000 light years away, it is visible to the unaided eye and is 20 times wider than the Orion Nebula.

This active region of star formation has enough gas to make a million stars. At its heart is one of the most enigmatic

*The whirling loops and superheated gas of the Carina Nebula epitomise starbirth in violent action.*

*In contrast, the serenity of the Lagoon Nebula reflects the softer side of star formation.*

and capricious stars known: Eta Carinae. It is a violent youngster tormented by the outbursts of stellar adolescence. In 1843, it erupted, becoming five million times brighter than our Sun. This unstable star, more than a hundred times heavier than the Sun, is destined to explode as a supernova.

### LAGOON NEBULA

Meanwhile, the Lagoon Nebula in the constellation of Sagittarius, is a picture of tranquillity. This gentle womb of burgeoning stars is home to so-called "Bok Globules," protostars encased in their natal dark dust.

# STAR CLUSTERS

After stars are born, they cling together like a clutch of nestlings. It is not until millions of years have passed that they fly the nest and make their own ways in space. In the beginning, the baby stars huddle together in a star cluster.

## THE PLEIADES

The best-known young star cluster is the Pleiades. The 19th-century British poet, Alfred, Lord Tennyson, described them in his epic *Locksley Hall* as "a swarm of fireflies tangled in a silver braid."

Although the cluster is well known as the Seven Sisters, skywatchers see any number of stars but seven. Most people can pick out the six brightest stars, while very keen-sighted observers can pick out up to eleven.

These are the most luminous in a group of at least 1,000 stars, lying about 400 light years away (although there's an ongoing debate about the precise distance). The brightest stars in the Pleiades are hot and blue, and all are young. Astronomers estimate that they are 75–100 million years old. They were all born together, and have yet to go their separate ways.

The fledgling stars have blundered into a cloud of gas in space, which looks like gossamer on photographs and in webcam images. Even to the unaided eye, or through binoculars, they are still a beautiful sight.

## SOUTHERN PLEIADES

Located in the constellation of Carina, this infant cluster — a twin to the Pleiades — contains about 60 stars. Nearly 500 million light years away, the cluster is a celestial baby; its stars are estimated to be a mere 50 million years old. Although visible to the unaided eye, the cluster is best viewed through binoculars, or with a telescope with a wide-angle eyepiece.

*Most famous of all star clusters, the Pleiades have drifted into an interstellar gas cloud — which gently reflects the young stars' blue color.*

## THE HYADES

Older star clusters lack the zest of their younger cousins. Their original blue-white stars are cooling to become red, and the clusters themselves are starting to drift apart.

The V-shaped Hyades star cluster, which forms the "head" of Taurus (the Bull), does not hold a candle to its neighbor, the dazzling Pleiades. But it is the nearest star cluster to the Earth, and it forms the first rung of the ladder in establishing the cosmic distance scale. By measuring the motion of the stars in the cluster, astronomers can establish their properties, and use these to find the distances to stars that are farther away.

Although Aldebaran, marking the bull's angry eye, looks as though it is part of the Hyades, this red giant just happens to be in the same direction, and it lies at less than half the distance. The Hyades cluster lies 153 light years away, and it contains about 700 stars. The stars, although aging, are all around 625 million years old.

## PRAESEPE

This swarm of stars in Cancer — also known as the Beehive Cluster — is a close twin of the Hyades in age, and it's moving in the same direction. The two clusters were probably born together.

You would be hard pressed to see Praesepe from a city, but if you are in a dark location, look between the constellations of Gemini and Leo to locate Cancer. Now concentrate your eyes on the central triangle of stars, then look inside.

Praesepe lies nearly 600 light years away, and contains over 1,000 stars, all of which were born together some 600 million years ago. Two of its stars are known to have planets in orbit about them, but they are not "Earths." Instead, they are "hot Jupiters," gas giants circling close in to their parent star.

Galileo, in 1610, was the first to recognize Praesepe as a star cluster, but the ancient Chinese astronomers obviously knew about it, as they had named it *Zei She Ge*, "the Exhalation of Piled-up Corpses!"

*Praesepe in Cancer is a more mature star cluster than the Pleiades. Its stars are less concentrated, and some red giants — evolved stars — have started to appear.*

# DOUBLE STARS

The Sun is an exception in having singleton status. Over half the stars you see in the sky are paired up: they are double stars (binaries) or even multiple stars. Somehow, they have never escaped the gravitational bonding of their birth and the close ties they forged in a star cluster.

You need look no further than the Plough (Big Dipper) for a beautiful naked-eye example. The penultimate star in the Plough's handle (at the bend) is clearly double.

Mizar (the brighter star), and its fainter companion Alcor, have been named "the horse and rider." Until recently, however, there was dispute as to whether they were really in orbit around one another, or just stars that happened to lie in the same direction. In the early 1990s, the sensitive Hipparcos satellite pinned down the distances of the pair to just over 80 light years. It seems that Mizar and Alcor are part of a six-member system, the other members of which are too faint to be seen by the unaided eye.

Another iconic double star that graces our heavens is Albireo — although astronomers are still discussing whether the star system is a genuine pair, or just two stars lined up. It lies in the constellation of Cygnus: a soaring swan, with his wings outspread as he flies along the Milky Way. Albireo is the end star in Cygnus, marking the swan's head.

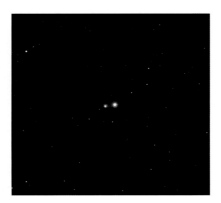

*Albireo, the "beak star" in Cygnus the Swan, is a beautiful contrasting double star, with yellow and blue components.*

*Mizar and Alcor, in the "handle" of The Plough (second from the left, here), are a striking naked eye double. The two are part of a multiple-star system.*

Use a small telescope, and you will be treated to one of the most glorious sky sights: a dazzling yellow star teamed up with a blue companion.

The yellow star is a giant, near the end of its life. It is 70 times bigger than the Sun, and 1,200 times brighter. The fainter blue companion is "only" 230 times as bright as the Sun.

The spectacular color contrast is due to the stars' different temperatures. The giant star is slightly cooler than our Sun, and shines with a yellowish glow. The smaller companion is far hotter: it is so incandescent, at 23,400°F (13,000°C), that it shines not merely white-hot, but blue-white.

# ECLIPSING BINARIES

Double stars dance with each other, performing a celestial tango. If the two stars are lined up, you can see the result. The brighter star is dimmed as its companion swings in front of it.

Algol, in the constellation of Perseus, is the classic "eclipsing binary." It represents the head of the dreadful Gorgon, Medusa. Persian astronomers clearly knew that there was something weird about this star. In Arabic, Algol translates as "The Winking Demon."

Watch Algol carefully and you'll see why. Every two days and twenty-one hours, Algol dims in brightness for several hours — to become as faint as the star lying next to it (Gorgonea Tertia).

In 1783, a young British amateur astronomer, John Goodricke of York, discovered Algol's regular changes, and proposed that Algol is orbited by a large dark planet that periodically blocks some of its light. We now know that Algol does indeed have a dim companion blocking its brilliant light, but that it is a fainter star, rather than a planet.

At the other end of the eclipsing binary star scale is Epsilon Aurigae (sometimes known as Almaaz), a star 130,000 times brighter than the Sun. It has recently been through an eclipse, in 2009–11, when it dropped to only half its normal brightness. This two-year event happens every twenty-

*Nineteen-year-old John Goodricke discovered Algol's changing brightness.*

*Algol, in the constellation of Perseus, marks the head of Medusa — the "Winking Demon."*

seven years. The reason for this extremely lengthy eclipse has baffled astronomers, as does the way Epsilon Aurigae usually brightens briefly at mid-eclipse.

Whatever the eclipsing object is, it must be huge: it would stretch beyond the orbit of Saturn if it were in our Solar System. The current thinking is that it is a dark disc of dust surrounding one or two stars that orbit Epsilon Aurigae itself, with their gravity acting as a "vacuum cleaner" to keep the central region clear.

# STARS IN THEIR PRIME

After the follies of a star's youth — restlessness, outbursts, and violent slingings of its material — our stellar neighbors settle down to become middle-aged, middle-class stars like our Sun. Welcome to the Main Sequence.

This is by far the longest part of a star's life, and it is pretty uneventful. Stars shine because they fuse hydrogen to helium in their cores, giving out energy. This balances the natural attraction of gravity, which wants to force the star to collapse.

But not all stars on the Main Sequence are the same. This is all to do with size or, more accurately, mass. They all power themselves by same nuclear reactions, but the more massive the star, the faster the reactions go and the quicker the end will come.

Heavyweight stars on the Main Sequence live fast and loose, and will explode after only a few million years. Middling stars, like our Sun, can hang around for billions of years. Low-mass stars can go on for trillions of years. Here are two stars, both in the prime of life, at the extremes.

## BROWN DWARFS

Gravity can create a ball of gas that's too lightweight for nuclear reactions to kick in, so it never shines as a star. Less than 8 percent of the mass of the Sun, these "failed stars" are called brown dwarfs, although the latest analysis suggests that, if we could see them close-up, they would look magenta in color.

### SPICA

Spica is huge; ten times heavier than the Sun, it is also 12,100 times more brilliant. Its surface is a searing at 40,300°F (22,400°C), compared to 9,900°F

(5,500°C) for our Sun. Before long, this giant star will explode as a supernova.

## Proxima Centauri

Proxima Centauri is our nearest stellar neighbor. This tiny red star, only one-seventh the size of our Sun, still packs a magnetic punch, with giant flares. But it's one-thousandth as bright as the Sun, and so energy-conservative that it will last another 400 trillion years.

*The Hubble Space Telescope homes in on Proxima Centauri, the nearest star to the Solar System.*

# EXTRASOLAR PLANETS

Our Sun is an average star, with its entourage of planets. What about other stars? Do they have planets?

The first discovery came in 1995, when Swiss astronomers Michel Mayor and Didier Queloz found that something was pulling the faint 51 Pegasi, near the great Square of Pegasus, backwards and forwards every four days. It had to be the work of a planet, tugging on its parent star.

Astonishingly, this planet is around the same size as the Solar System's giant, Jupiter, but it is far closer to its star than Mercury. Astronomers call such planets "hot Jupiters."

A team in California led by Geoff Marcy was already looking for planets, and soon found more. Now astronomers are finding whole solar systems with several worlds in stable orbits, like the five planets orbiting the star Kepler-186. One of these worlds is in the "Goldilocks zone," where conditions are not too hot, and not too cold, for liquid water. It is possibly a rocky planet not much

*Artist's impression of the young planetary system surrounding Beta Pictoris. The star is surrounded by a dusty disc, thought to be forming into planets.*

more massive than Earth, which may harbor liquid water.

The tally of planets circling other stars now stands at around 2,000, with many more waiting to be confirmed

The latest breakthroughs were made by NASA's Kepler orbiting spacecraft, which picked up tiny diminutions in light when a planet crossed the disc of its parent star. Kepler researchers believe that the data they have garnered so far point to "at least" 17 billion Earth-sized planets in our galaxy. Many of these could be the abode for life.

*Planet-hunter Geoff Marcy contemplates the heavens.*

## INSIDER'S GUIDE TO THE TOP 10 EXTRASOLAR PLANETS

| NAME | DISTANCE (LIGHT YEARS) | LENGTH OF "YEAR" | MASS COMPARED TO EARTH | TEMPERATURE | TYPE |
|------|------------------------|------------------|------------------------|-------------|------|
| 51 Pegasi b | 51 | 4.2 days | 150 | 1,820°F (1,010°C) | Hot Jupiter |
| PSR B1257 c | 1,000 | 67 days | 4.3 | Not known | Pulsar planet |
| PSO J318 | 80 | Not known | 2,000 | 1,600°F (900°C) | Orphan planet |
| TrES-2 b | 750 | 2.5 days | 380 | 1,800°F (1,000°C) | Black planet |
| Kepler-78 b | 400 | 8.5 hours | 1.8 | 4,500°F (2,500°C) | Molten Earth |
| OGLE-2005-BLG-390L | 21,000 | 9.5 years | 6.7 | -360°F (-220°C) | Frozen Earth |
| Kepler-16 b | 200 | 229 days | 105 | -130°F (-90°C) | Twin suns |
| Fomalhaut b | 25 | 2,000 years | 100? | -330°F (-200°C) | Longest year |
| Kepler-37 b | 215 | 13 days | 0.01 | 800°F (430°C) | Smallest planet |
| Gliese 667C c | 22 | 28 days | 6 | 40°F (4°C) | Most Earth-like |

# LIFE ON OTHER PLANETS

Whenever we give a presentation, we are always asked, "Is there anybody out there?" It is now half a century since the veteran American astronomer Frank Drake turned his radio telescope to the heavens in the hope of hearing an alien broadcast. Despite false alarms (probably caused by secret military equipment), there has been a deafening silence.

Drake and his colleagues founded an independent institution in California, the SETI Institute (Search for Extra-Terrestrial Intelligence). It is a serious scientific endeavor that that not only looks for signals, but explores the possible biology and psychology of alien life.

Recently, their fortunes have been boosted by a donation from Microsoft co-founder Paul Allen. Thanks to Allen, the team is now building an array of 400 radio telescopes in California to tune into that first whisper from ET.

*Young Frank Drake (center), with two of his SETI-detecting team*

*Part of the Allen Array of radio telescopes in California, dedicated to SETI*

Is life out there far more advanced than us? Is radio communication something that came and went? The SETI researchers are contemplating communication by laser beams, but even that may prove too primitive.

Why are astronomers so optimistic that there could be alien life? It is partly a result of findings on our own world —the discovery that life can survive in the most extreme environments. No, we are not talking about little green men; it is more about little green slime.

Bacteria called "extremophiles" can exist in the most bizarre places — the deep ocean trenches, the hearts of nuclear reactors and the vacuum of space.

Moving out to the wider arena of the Solar System, there are many more potential habitats for primitive life. Mars, almost certainly, has bugs. Jupiter's ice-coated moon Europa is thought to harbor an ocean in which alien fish can swim.

Moving farther afield, *could* there be intelligent life on the thousands of worlds beyond our Solar System?

It's up to you to find out here! Log on to SETI @home, or SETI-live (see page 393), to search for data from radio telescopes that are looking for alien communications. You might just pick up THE signal!

# RED GIANTS

Middle-aged spread is a fate that not only befalls humans; it also happens to stars (although the causes are rather different). Stars like the Sun generate energy by nuclear fusion; they "burn" hydrogen into the next element, helium, in their hot cores.

But there is only a finite supply of hydrogen in a star's powerhouse. When it runs out, after millions or billions of years — depending on the star's mass — the core, now made of helium, collapses to a smaller size and begins fusing helium into carbon.

These reactions create more heat. But this, ironically, causes the outer layers of the star to balloon in size and, consequently, cool down. Gone is the star's golden glow. It is replaced by the baleful red or orange color of a red giant star.

Our heavens boast some glorious red giants. For instance, brilliant Arcturus always gladdens our hearts as a harbinger of summer. But alas, its orange color indicates that it is near the end of its life.

Betelgeuse, in Orion, is one of the biggest stars known. If placed in the Solar System, it would swamp the planets all the way out to Jupiter. And it is one of just a few stars to be imaged as a visible disc from Earth. Around 1,000 times wider than the Sun, Betelgeuse falls into the class of supergiants. It also fluctuates slightly in brightness as it tries to get a grip on its billowing gases.

*Red supergiant star Betelgeuse*

*Giant Antares, the heart of Scorpius. Two globular clusters join the star in this frame: the smaller NGC 6144 to the right of Antares, and the larger M4 just below center.*

One of the reddest red giants in the sky is Antares. Its name means "the rival of Mars" — and you can see why. It is a ruddy red star that marks the heart of the constellation of Scorpius. Residing some 550 light years away, Antares is a bloated star that is at least 15 times heavier than our Sun. Now it has grown 880 times bigger then our local star and 57,500 times more luminous.

Placed at the center of our Solar System, it would reach out to the asteroid belt. However, its size is not constant. Antares' gravity cannot cope with its extended girth, making the star swell and shrink, changing in brightness as it does so. The giant star has a small blue-white companion, which is hard to see against Antares' glare. Just visible in a small telescope, the star circles the red giant every 878 years.

# VARIABLE STARS

Like human beings, stars can swell and shrink. It's especially true of older stars, which have experienced middle-age spread. As their outer layers shrink in and swell out, they change in brightness.

The most iconic variable star in the sky is Delta Cephei, in the constellation representing King Cepheus from Greek mythology. Hardly worthy of special attention, it is a yellowish star of magnitude four, just visible to the naked eye, but not prominent. A telescope reveals a companion star. But this star holds the key to the size of the Universe.

Check the brightness of this giant star carefully over days and weeks, and you'll see that its brightness changes regularly, every 5 days and 9 hours. This is a result of the star literally swelling and shrinking in size, from 32 to 35 times the Sun's diameter.

Astronomers have found that stars like this, known as Cepheid variables, show a link between their period of variation and their intrinsic luminosity. By observing the star's period and brightness as it appears in the sky, astronomers can work out a Cepheid's distance.

Essentially, Cepheid variable stars have turned out to be cosmic measuring rods. Now, with the Hubble Space Telescope, astronomers have measured Cepheids in the Virgo Cluster of galaxies, which lies 55 million light years away.

While Cepheids are reliable and

*Henrietta Leavitt studied Cepheids (also see page 259), helping to discover the distance scale of the Universe*

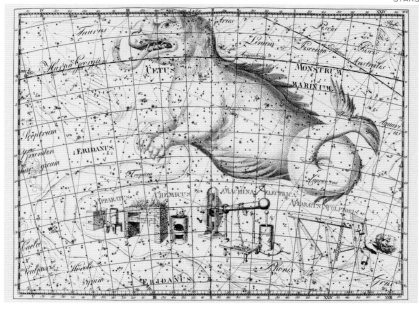

*Mira, in Cetus the Whale, is a variable star — but not as regular as a Cepheid. An ancient red giant, it billows in and out, changing in brightness unpredictably.*

predictable, other variable stars are more eccentric. Take Mira in the constellation of the Whale, Cetus. It was first observed in 1596 by the German astronomer David Fabricius. He thought that it was an exploding star, because it flared up and then vanished. It was back 332 days later.

Fabricius had discovered a star close to the end of its life, which he called "the wonderful" (*Mira* in Latin). It is so vast

and swollen that, were it to be placed in the Solar System, it would engulf all the planets out to Mars.

Gravity just can't get a grip on this star. So, it balloons in and out, changing in brightness from magnitude two (comparable to the Pole Star) to magnitude 10. It might be time to bring out the telescope.

# PLANETARY NEBULAE

William Herschel — who discovered the planet Uranus — first named these fuzzy objects "planetary nebulae," because, to him, they looked like the planet that he had found. Now we know that they are the end-point for a star like the Sun.

When a star like the Sun runs out of its nuclear fuel at the end of its life, its core contracts, heating up its outer layers. The star becomes unstable and puffs off its atmosphere into space in the shape of a ring.

The ring disperses in a few thousand years. It leaves the derelict core inside to leak away its energy, as a white dwarf star that will eventually cool to end up as a cold, lonely black dwarf.

Tucked into the small constellation of Lyra (the Lyre), near the brilliant star Vega, lies a strange celestial sight. It was first spotted by French astronomer Antoine Darquier in 1779, as "a very dull nebula, but perfectly outlined; as large as Jupiter and looks like a fading planet." Under higher magnification, it appears as a bright ring of light with a dimmer center, so it is known as the Ring Nebula.

You can make out the Ring Nebula with even a small telescope, although it is so compact that you will need a magnification of over 50x to distinguish it from a star.

The tiny, faint constellation of Vulpecula ("the Little Fox") contains one of the most beautiful sights in the sky: the Dumbbell Nebula. At magnitude 7.5, it is just visible through binoculars, and a

*Small but perfectly-formed stellar wraith: the Ring Nebula*

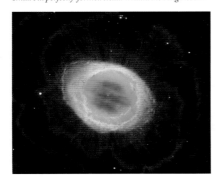

real treat when seen through a telescope. Although an object of beauty, the Dumbbell is doomed.

The Dumbbell lies roughly 1,360 light years away (distances to planetary nebulae are notoriously difficult to measure), and is an estimated 2.5 light years across. From the expansion rate of the nebula, astronomers are able to estimate that this star died some 10,000 years ago.

*The Dumbbell planetary nebula shed its outer envelope of gas some 10,000 years ago. It is a fate that will befall our Sun.*

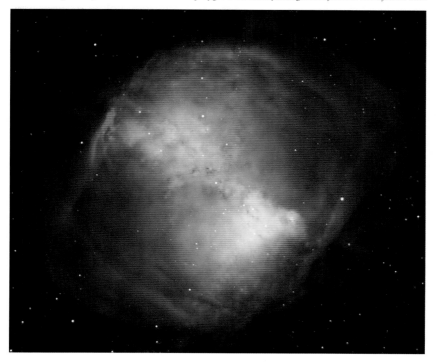

# WHITE DWARFS

Sirius is the brightest star in the sky, shining at a magnitude of -1.47. It isn't a particularly luminous star; it just happens to lie nearby, at a distance of 8.58 light years.

The "Dog Star" is accompanied by a very dim companion, which is affectionately called "The Pup." American astronomer and telescope maker Alvan Clark discovered this tiny star in 1862 when he was testing a telescope, but its presence had been predicted by the German astronomer Friedrich Bessel nearly 20 years before, when he had observed that something was "tugging" on Sirius. The Pup is a white dwarf — the dying nuclear reactor of an ancient star that has puffed off its atmosphere as a planetary nebula (see page 238–9), now long since faded away.

White dwarfs are the size of a planet, but have the mass of a star: because they are so collapsed and dense, they have considerable gravitational power, which explains Sirius' wobble. The Pup is visible through medium-powered telescopes.

## NOVAE

Occasionally a "new star" — a nova, Latin for "new" — surprises us in the skies. This is all down to white dwarf stars. Most stars are double, and in pairs where one star is a white dwarf, the diminutive but powerful beast pulls on its companion, dragging gas over to its own surface. Suddenly, the accumulated gas ignites, like an uncontrolled cosmic hydrogen bomb. For weeks, the nova can shine 100,000 times brighter than the Sun.

*The "Firework Nebula" — a cloud of expanding gas surrounding Nova Persei, which exploded in 1901. The central star is a white dwarf.*

In seven billion years time, our Sun will become a white dwarf, when it runs out of hydrogen fuel. A white dwarf star is one of the triumphs of gravity. What was once a powerful nuclear reactor at the star's core, collapses, having run out of its hydrogen fuel. The core becomes so crushed that atoms themselves cannot stand the strain; they break up into electrons and nuclei, producing a density far higher than we can create on Earth.

In fact, a single matchbox full of material from a white dwarf would weigh as much as an elephant.

What's the future for a white dwarf? The answer is one of slow decay. The dying star's fate is to become a cold, black cinder.

# SUPERNOVA!

In November 1572, the Danish nobleman and astronomer Tycho Brahe was amazed to see a brilliant star in the constellation Cassiopeia, where none had existed before. It shone as brightly as the planet Venus, then gradually faded from sight.

Tycho called it a nova, meaning "new star." Now, however, we would describe this apparition as a supernova: the spectacular suicide of a star that has blown itself apart in its death throes.

There are two kinds of supernova. Tycho was observing the explosion of a tiny white dwarf star (see page 240), as what astronomers call a Type Ia supernova. When a giant star detonates it is denoted as being Type II (see pages 244–5).

Unknown to Tycho, a white dwarf had been scooping up gas from a close companion star. But white dwarf stars have a natural weight limit. When it became 40 percent heavier than the Sun, the tiny star became unstable. Carbon and oxygen within the white dwarf fuelled a colossal nuclear explosion that entirely destroyed the tiny star. Its funeral pyre briefly burnt as brilliantly as billions of Suns.

Today, astronomers regularly scan the skies for Type Ia supernovae in remote galaxies, because they provide a key to measuring the galaxies' distance, and from that, the size and age of the Universe.

*Tycho Brahe, pointing to the position of his supernova of 1572*

*More than 400 years later, the debris from Tycho's supernova lights up space in an incandescent fireball.*

## AMATEUR SUPERNOVA SLEUTHS

Most supernovae in distant galaxies are discovered by automated telescopes run by professional astronomers, but there is still a role for the dedicated backyard stargazer.

The Reverend Robert Evans, in New South Wales, Australia, has discovered an amazing 42 supernovae by regularly looking, by eye, at galaxies through his telescope. He remembers the normal appearance of 1,500 galaxies, and can immediately notice any unexpected star.

In the UK, Tom Boles is another amateur astronomer, who uses electronic equipment to pick out new stars in familiar galaxies. He has found (as of 2014) an incredible 155 supernovae.

# SUICIDE OF A GIANT STAR

Supernovae, exploding stars, have been logged since ancient times. The Chinese called them "guest stars." Apart from exploding white dwarfs (see page 242), they are stars that die young and violently, with all the repercussions this will have on their planets and any life forms.

Stars more than eight times more massive than the Sun are doomed to an early death. They rip through the nuclear reactions that power them at a reckless rate. While our modest Sun converts hydrogen to helium in its core, the supergiants are far more ambitious.

When the hydrogen, and then the helium, run out, their gravity compresses the core tighter, building successively new elements in the star's central nuclear reactor, while the outer layers billow out to become a red giant. All goes well, and the star continues shining, until the core is made of iron.

Trying to fuse iron is a fatal mistake. Iron fusion takes *in* energy and, as a result, the core catastrophically collapses. The star cannot stand up to the shock. A burst of neutrinos blasts through its outer layers, blowing it apart.

At its maximum brightness, this kind of supernova (Type II) can outshine a whole galaxy of 100,000 million stars. The most spectacular recent supernova was 1987A, in the Large Magellanic Cloud. Easily visible to the unaided eye, it was 250 million times more luminous than the Sun.

The aftermath of this violent death is a tangled fireball of gases. The supernova hurls a cornucopia of elements into space, those from inside the dead star and others created in the inferno of the explosion. The gold in a wedding ring was born in a supernova explosion.

In the end, a supernova is a phoenix, for out of its ashes, the seeds of life — elements like carbon — will arise, to be scattered into space. We owe our very own existence to supernovae.

*In 1987, a supernova exploded in the Large Magellanic Cloud, a nearby galaxy. This is what remains.*

# PULSARS

Home in on the constellation of Taurus with a small telescope, and focus in on a small region above the bull's southern "horn." There, in 1054, Chinese astronomers witnessed the appearance of a "new star," which outshone all the other stars in the sky. It was visible in daylight for 23 days and remained in the night sky for nearly two years. But this was no new star. It was an old star on the way out, which

*Convoluted filaments of the Crab Nebula, the remains of a star that exploded in 1054 AD.*

exploded as a supernova, because it was overweight.

Today, we see the remnants of the star as the Crab Nebula, so named by the 19th-century Irish astronomer Lord Rosse, because it resembled a crab's pincers. Even today, the debris is still expanding from the wreckage, and it now measures 15 light years across.

At the center of the Crab Nebula is the core of the dead star, which has collapsed to become a pulsar. This tiny, but super-dense object—only the size of a city but with the mass of the Sun—is spinning around furiously at 30 times a second and emitting beams of radiation like a lighthouse.

Pulsars were discovered in 1967 in Cambridge,UK, by the young graduate student Jocelyn Bell. All summer, she had been building a very unconventional radio telescope, made of 1,000 wooden posts and 190 km of wire. And in November that year, she started picking up signals from an

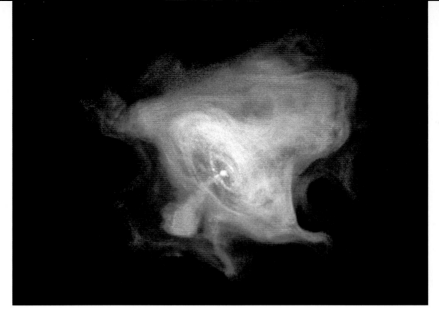

*Chandra X-ray satellite image of the center of the Crab Nebula, showing its energy-generating pulsar*

object that was flashing radio waves every 1.337 seconds, as regular as a clock.

At first, Bell's colleagues wondered if they were receiving a deliberate signal from ET and nicknamed it LGM-1 ("Little Green Man 1").

Other scientists were on the case. They worked out that a massive supernova explosion would leave the dying star's core behind as a neutron star, an object so compacted by the collapse of the former star that a pinhead of its matter would weigh a million tons. These stars are so compressed that they are made entirely of neutrons.

Spinning neutron stars, pulsars, are highly magnetic. That is why they appear to pulse: they emit jets of electromagnetic radiation from their poles, which regularly sweep past us.

We have discovered over a thousand pulsars. But none of them will carry on forever. As time passes, a pulsar's spin-rate will decrease, and its beat will stop.

# BLACK HOLES

In 1970, space scientists launched a new satellite into the cosmos. Its mission was to target objects emitting powerful X-rays, radiation that does not penetrate our atmosphere.

X-rays are a sign of violence in the Universe. And the Uhuru satellite made a big find. Cygnus X-1 is a double star system. The primary star is a blue supergiant weighing in at 30 times the mass of the Sun. But it is upstaged by its companion: an object with 10 times the Sun's mass.

The companion is literally devouring the supergiant. Its vast gravity is tearing matter away from the large star, creating immense energy, and generating X-rays that are almost off the scale.

Welcome to the world of black holes.

A black hole is the ultimate star corpse and one of the most exciting new discoveries in astrophysics. However, their existence was theorized in 1783 by Reverend John Michell, the vicar of the village of Thornhill, in Yorkshire in the north of England. He realized that the most massive objects in the Universe would be invisible, because of the pull of their gravity.

Black holes are star death stretched to the limit. When a star as much as 20 or 30 times as massive as the Sun explodes in a supernova, its core collapses so

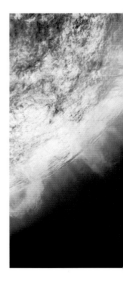

*Artist's impression of the first black hole to be discovered, Cygnus X-1. The black hole is tearing matter off its blue supergiant companion star. The gas orbits the cosmic predator in a frenzied, fiercely-glowing accretion disc — before disappearing into the hole, and from our Universe, forever.*

dramatically that its immense gravity prevents light from escaping. Its gravitational pull also sucks objects into its abyss, from which they can never return, because nothing can travel faster than light.

On a cosmic scale, black holes are tiny, at no more than 30 km across, but they pack a massive punch. If they are in altercation with another star, they attract its gas relentlessly, forming a swirling vortex around the hole, called an accretion disc. As the gas falls in, it moves faster and faster until it is moving at close to the speed of light.

Friction makes the gas ferociously hot, and the accretion disc glares brilliantly in X-ray radiation.

The big question is: where does the matter that falls into the black hole go? Some scientists believe that it enters another universe. In which case, black holes could be gateways to a new cosmos.

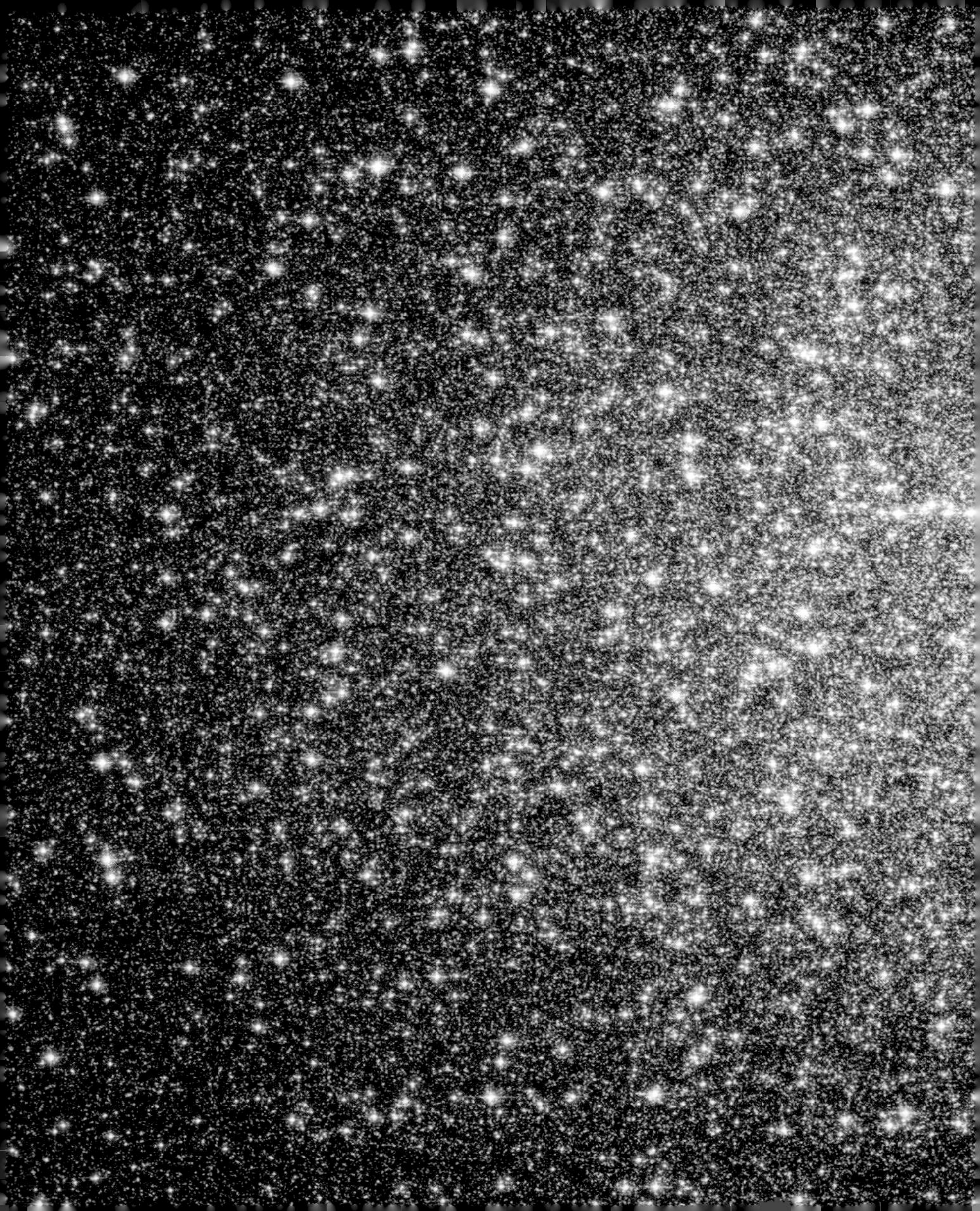

# INTRODUCTION

On the grand scale, our cosmos is teeming with billions of galaxies. These star cities are hosts to billions of stars, and each has its own character.

You can see our own galaxy by finding a dark site to view the night sky. The stars of the Milky Way reach out overhead: a vista of our distant stellar companions in the far-reaches of its Catherine-wheel-shaped disc.

Every galaxy is different. There are elegant, spiral-shaped galaxies, like the Milky Way. These are young, active galaxies, busy making new stars. At the other end of the spectrum are the elliptical galaxies. These objects have reached the cosmic menopause: devoid of gas and dust, they are past their star-forming days.

Last, there are the active galaxies, the quasars and radio galaxies, which harbor supermassive black holes in their cores, causing cosmic mayhem.

What galaxies have in common is that, in general, they are receding from each other. The Universe is expanding, as a result of its birth in the Big Bang 13.8 billion years ago. The expansion is poised to continue, and even to accelerate. Eventually, the Universe will become an empty, lonely place.

*On its largest scales, our Universe is populated by magnificent denizens of the cosmos, such as this glorious spiral galaxy NGC 1232 in Eridanus. This star city contains billions of stars.*

# THE MILKY WAY — AN INSIDER'S GUIDE

Arching over our heavens, the gently glowing band of the Milky Way is an awesome sight under dark, transparent skies. The name is a translation of the Latin *Via Lactea,* commemorating the ancient myth that it was a stream of milk gushing from the breast of the goddess Juno, as she suckled the infant Hercules.

But every culture has its own interpretation. Native North Americans saw the Milky Way as the path of ghosts heading for the afterlife. To Inuit, it was a trail of ashes that led travelers home.

Galileo was the first to work out its true nature. He applied his telescope to the sky. And with sparse magnification, he observed that the Milky Way was made up

of "a congeries of innumerable stars."

Look at the Milky Way through even a modest pair of binoculars, and you will have a view much like Galileo's. The continuous band breaks up into myriad distant stars, looking as though they are closely packed together. Follow the band across the sky, and you will pick out star clusters and nebulae as you travel its length.

These stars, clusters and nebulae are all denizens of our local galaxy, flattened into a band because we are viewing a disc of stars, edge-on. It is akin to seeing the overlapping streetlights of a distant city on Earth.

The Milky Way is home to over 200 billion stars, with the Sun about halfway out. The center of the Milky Way lies in the direction of the constellation Sagittarius, but our view of it is obscured by great clouds of dark dust that block the view for even the most powerful telescopes.

*The distant stars of our local Galaxy, the Milky Way, arch across the skies of Earth. The black swathe across the band is interstellar dust — building block of future stars and planets.*

# ANATOMY OF OUR GALAXY

If you were able to soar above the Milky Way, you would gaze down on a glorious cosmic Catherine wheel, clasped in beautiful spiral arms.

The Milky Way is surprisingly thin. Apart from the thicker central regions, our galaxy has the same proportions as a pair of CDs placed on top of each other: 100,000 light years wide, but only 2,000 light years thick. The spiral arms are regions where stars are more closely packed, along with interstellar gas and dust.

Where the interstellar matter is squeezed within the spiral arms, it is condensing into new generations of stars and planets. The vigorous young stars light up their surroundings, so the spiral arms are studded with glowing red nebulae.

## GLOBULAR CLUSTERS

While the spiral arms host the youngest stars in the Milky Way, the oldest inhabitants reside in the center, a straight "bar shape" of ancient stars, or in 150 giant clusters of stars, scattered all around the galaxy.

These globular clusters typically contain about a million stars each. They appear in the sky as round fuzzy patches. Several are visible to the naked eye: Omega Centauri and 47 Tucanae in the south and M13 in the northern constellation of Hercules.

Seek out these galactic fossils, and you are seeing stars born just after the formation of the Milky Way itself, over 13 billion years ago.

*Bird's-eye view of the Milky Way, assembled from data captured by the Spitzer Space Telescope. Its nucleus is bar-shaped, encoiled within two spiral arms.*

The galaxy's main features include the Sagittarius Arm and the Scutum-Centaurus Arm. The Sun lies in a branch of the Perseus Arm, called the Orion Spur — as its name suggests, what would draw your eye to this region is the great Orion Nebula and its entourage of brilliant stars. The Sun itself would be lost in the midst of thousands of other average stars.

# OUR GALAXY'S COMPANIONS

The Milky Way has an entourage of smaller companions swarming around it. Most of these are small, scruffy irregular galaxies, made up of just a few million stars. Many of them have had gravitational skirmishes with our galaxy in the past, and have come off by far the worst. However, there are two exceptions, both of which are glorious naked-eye sights south of the Equator. These are the Large and Small Magellanic Clouds (LMC and SMC), two irregular galaxies that circle our Milky Way

*Young stars recently born in the Small Magellanic Cloud pour out their radiation as X-rays — a sign that their magnetic fields are very active.*

## MEASURING THE UNIVERSE

In 1908, Henrietta Leavitt (see page 236), a young assistant at Harvard College Observatory, was looking at images of variable stars in the Small Magellanic Cloud. She realized that their variation followed a pattern: the brighter the star, the longer it took to change in brightness.

The stars were all in the same community. Could their variations be used to measure the distance to the Small Magellanic Cloud?

These particular stars were called Cepheid variables: yellow giants that billow in and out. If astronomers could find the distance to a nearby Cepheid in the Milky Way, they would have a stellar measuring rod.

Soon, they pinned the distances down, and Leavitt established how far away her Cepheids in the SMC were. Her Cepheid technique was the breakthrough in finding distances to other galaxies and working out the scale of the Universe.

at distances of roughly 160,000 and 200,000 light years, respectively.

These are substantial small galaxies. The LMC measures 14,000 light years in diameter; the SMC comes in at half its size. Both have been disrupted by altercations with the Milky Way; in fact, the SMC has practically been torn apart.

The LMC is frenetically active in making stars. It hosts a gigantic stellar nursery, the Tarantula Nebula, which is 300 light years across. If it were placed at the distance of the Orion Nebula, it would cast shadows on Earth.

Although both the LMC and SMC are classified as irregular galaxies, there is a lot of evidence from their structure that they are disrupted barred spiral galaxies each with a rectangular "bar" of old stars crossing its nucleus.

# SPIRAL GALAXIES

Spiral galaxies, like our own Milky Way, are some of the most fabulous occupants of our Universe. And we are blessed with one that can be seen with the unaided eyes.

### ANDROMEDA GALAXY

The Andromeda Galaxy covers an area four times bigger than the Full Moon. Like our Milky Way, it is a beautiful spiral shape, but, sadly, it is presented to us almost edge-on.

Our intergalactic neighbor lies about 2.5 million light years away, and it is similar in size and shape to the Milky Way. It also hosts two bright companion galaxies and, just like our own star city, a flotilla of orbiting dwarf galaxies.

Unlike other galaxies, which are receding, the Milky Way and Andromeda

*The Andromeda Galaxy — closest major spiral to our Milky Way. Visible to the unaided eye, it is home to a billion stars.*

*The smaller Triangulum galaxy lies close to Andromeda in the sky. The Milky Way, Andromeda and Triangulum are all members of the Local Group of galaxies*

are in fact approaching each other. Astronomers estimate that the two will merge in about five billion years' time. The result of the collision may be a giant elliptical galaxy, nicknamed Milkomeda, which will be devoid of the gas that gives birth to stars, and dominated by ancient red giants.

### TRIANGULUM GALAXY

The next-nearest spiral is the Triangulum Galaxy, which lies close to Andromeda in the heavens. If you have a *very* dark sky, it is just visible to the unaided eyes.

This galaxy lies about three million light years away. It is only half the size of the Milky Way, but is face-on, so we get to see its spiral arms in all their glory. It is a fanatic at star formation and boasts a starbirth factory 60 times larger than the Orion Nebula.

## SCULPTOR GALAXY

This fabulous spiral galaxy is visible through binoculars, although a telescope helps. Nicknamed the "Silver Coin," it lies in the obscure constellation of Sculptor.

It was discovered by Caroline Herschel, whose brother William discovered the planet Uranus in 1781.

Like the Andromeda Galaxy, the Sculptor Galaxy (officially known as NGC 253) lies almost edge-on to us. Although we do not have a view of its spiral arms, we have discovered a great deal about the activity in its galactic center.

It is undergoing a period of intense

*Southern-hemisphere jewels: NGC 253 in Sculptor. A great target for owners of a telescope more than 300 mm across*

*The face-on spiral galaxy M83 in Hydra is a hotbed for starbirth and star death and home to 200 trillion suns*

star formation, possibly as a result of a brush with a dwarf galaxy that collided with it 200 million years ago. The crash stirred up the Silver Coin, making it into a "starburst galaxy."

### SOUTHERN PINWHEEL

Visually, this stunning galaxy couldn't be more different from the Silver Coin. Also visible through binoculars, the "Southern Pinwheel" in Hydra is completely in-your-face. Catalogued as M83, it flaunts its glorious spiral arms to excess. One of the most beautiful spiral galaxies in the sky, it is a factory for both star birth and star death. Six supernovae (exploding stars) have been logged in the Southern Pinwheel since 1923.

# ELLIPTICAL GALAXIES

Elliptical galaxies come in two sizes: large and small. What they all have in common is that they are made up of old, red stars and are bereft of the gas, so common in spiral galaxies, from which new stars form.

Andromeda's two major attendants are both dwarf elliptical galaxies. M32 measures a mere six million light years across and is very compact. It is made up of aging red and yellow stars and, with no interstellar gas as fuel, it is a galaxy on the way out. However, it does have a massive black hole at its core, which could stir up things in the future.

NGC 205, Andromeda's other dwarf elliptical companion, is, like its neighbor, visible through a small telescope. Unlike most dwarf elliptical galaxies, NGC 205 seems to have some fire in its belly — there are hints of recent star formation.

The big ellipticals are in a different category altogether. M87 in Virgo hosts 6,000 billion stars. It harbors a central black hole, which astronomers estimate measures four billion times the mass of our Sun making it one of the biggest black holes known. The black hole has caused such mayhem that M87 shoots out a jet of gas 5,000 light years in length, traveling at a speed close to the velocity of light.

South of the Equator, Centaurus A is busily devouring a smaller spiral galaxy. This giant elliptical, 1,000 billion times the mass of our Sun, is easily visible with

*A powerful jet emerging from its central supermassive black hole dominates M87.*

*Centaurus A is a southern-hemisphere sensation. Here, the elliptical monster is devouring its spiral companion.*

binoculars. It has a gash of thick dust crossing its center: the remnants of the galaxy that it is feasting on. It boasts a black hole weighing in at 55 million Suns.

Like M87, the galaxy also hurls jets of gas into space, in this case, over a million light years long.

# GALAXIES IN EMBRACE

Galaxies like to dance with each other. And the results of their choreography are often sensational. Pairs, or trios, of galaxies swirl around in the darkness of the cosmos, extracting exquisite filaments of glowing matter from each other.

Of all the interacting galaxies, the Whirlpool is the most iconic. This beautiful face-on spiral galaxy (catalogued as M51), 23 million light years away, is interacting with its smaller companion, NGC 5195 and connected to it by a bridge of stars and gas. Computer simulations

*The glorious Whirlpool — this cosmic dance has invigorated the spiral into a frenzy of starbirth.*

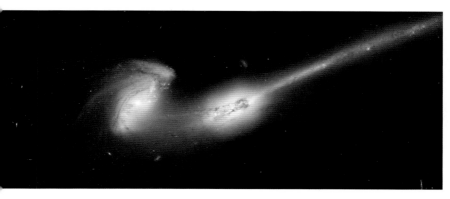

*"The Mice" duo has created an outburst of star formation — mouse-like tails of stars and gas.*

show that NGC 5195 passed through the main disc of the Whirlpool about 500 million years ago, stimulating starbirth in the larger galaxy, and creating its elegant spiral structure.

Slightly farther away, at 45 million light years, is a pair of interacting galaxies called "the Antennae." Situated in the constellation of Corvus, these were once separate galaxies. Now they are an (albeit beautiful) celestial mess. The two galaxies, both of them spirals, merged 600 million years ago.

The impact ejected stars from both galaxies, producing the pair of long stellar streamers on either side of the merged galaxy that give it its nickname. The new galaxy is undergoing an outburst of star formation, as the gas clouds from the two galaxies collide.

Not yet in collision, but close to it, are "the Mice" in Coma Berenices. Nearly 300 million light years away, the pair started to get close to each other around 290 million years ago. Their interaction has created a long, mouse-like tail of stars.

The culprit is galactic tides: each galaxy's gravity is stretching and disrupting the other. Like the Antennae, the Mice will eventually merge.

# ACTIVE GALAXIES AND QUASARS

Last century, when astronomers started using radio telescopes to survey the sky, like the giant dishes at Jodrell Bank in the UK, and Arecibo in Puerto Rico, they discovered mysterious sources originally thought to be "radio stars." On close inspection with optical telescopes, the "stars" turned out to be giant elliptical galaxies.

One of the most powerful, and among the brightest, radio sources in the sky is Cygnus A. It is a radio galaxy 600 million light years away, and it certainly packs a punch. The galaxy emits jets from its tortured core that stretch over half a million light years into space, billowing into gigantic clouds when they hit the intergalactic medium — the tenuous matter between the galaxies.

Radio galaxies are one thing, but quasars are quite another. In the early 1960s, a young Dutch astronomer, Maarten Schmidt, had the bright idea of taking the spectrum of a "radio star" called 3C 273.

When he split up the light patterns, he discovered that 3C 273 was rushing away from us at a colossal speed, as a result of the expansion of the Universe. This meant that the quasar had to be both incredibly distant and awesomely powerful.

Schmidt's measurements put 3C 273 at a distance of 2.5 billion light years, and this discovery was just the beginning. Of the 200,000 quasars known to exist today,

*Maarten Schmidt contemplates the spectrum of a quasar.*

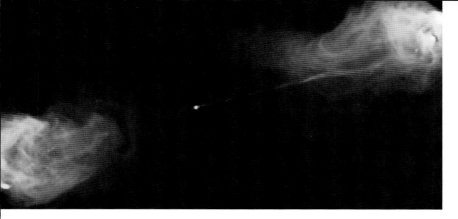

*Radio galaxy Cygnus A is fuelled by a supermassive black hole weighing 2.5 billion suns at its core. Jets, traveling close to the speed of light, create two enormous gas clouds in interstellar space.*

the farthest lies a staggering 30 billion light years away!

Quasars are baby galaxies in their disruptive birth throes. At the core of each lurks a supermassive black hole, gobbling stars and gas. Their fetal burps cause massive explosions in the Universe. The black hole at the core of 3C 273 weighs in at a billion times the mass of our Sun.

Even the Milky Way boasts a mini-quasar. Infrared telescopes can penetrate the dust that shrouds it to see the heat radiation from stars and gas clouds at the galactic center. These objects are speeding around so fast that they must be in the grip of something with fantastically strong gravity. One star circles the Galaxy's core at over 18 million kph. Radio astronomers have found that the Galaxy's exact heart is marked by a tiny source of radiation: Sagittarius A*.

Putting all these observations together, astronomers have concluded that the heart of our Milky Way must contain a supermassive black hole. They estimate that it weighs in at four million Suns.

When a speeding star comes too close to this invisible monster, it is ripped apart. There's a final outburst from the star's gases, producing the observed radio waves, before it falls into the black hole and disappears from our Universe.

# GEOGRAPHY OF THE COSMOS

Galaxies are gregarious creatures. They like living in swarms, bound together by the ties of gravity. The structure of the Universe is dictated by galaxy clusters: groups of star cities whose members interact with each other.

Our own Milky Way is no exception. It is a member of a small cluster of over 50 galaxies, known as the Local Group. The Alpha males in the pack are the Andromeda Galaxy and our Milky Way, with the Triangulum Galaxy coming up third. The rest of the Local Group is a mixture of dwarf elliptical and irregular galaxies.

Our own group is just a part of a mega-construct on the cosmic scale: the Local Supercluster. Our humble bundle of galaxies, along with others, plays gravitational tag with the massive Virgo Cluster, which forms the core of our local celestial neighborhood and lies about 55 million light years away.

If you have a small telescope, sweep the "bowl" formed by the constellation Virgo's "Y" shape, and you will detect dozens of fuzzy blobs. These are just a handful of the thousands of galaxies making up the Virgo Cluster.

Many of the 2,000 galaxies in the Virgo Cluster are spirals like our Milky Way, but some are even more spectacular. The heavyweight galaxy of the cluster is M87, a giant elliptical galaxy emitting a jet of gas 5,000 light years long at almost the speed of light.

On the largest scale, superclusters like Virgo are key to the architecture of the Universe. The cosmos, on its widest scale, resembles an enormous Swiss cheese, full of holes. Filaments of galaxy superclusters measuring more than 10 billion light years across are the most massive features in the cosmos. They surround immense empty voids that are bereft of galaxies.

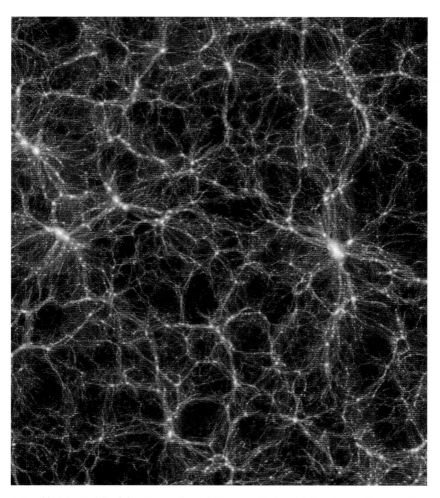

*A view of the Universe a billion light years across shows galaxies arranged in elongated filaments around immense voids.*

# THE BIG PICTURE

In the beginning was the Big Bang — 13.8 billion years ago, our Universe was created in a fury of fire and frenzy. No one knows where the Big Bang came from, or why it happened.

The pointers to the origin of our Universe were first noticed by the American astronomer Edwin Hubble, observing galaxies in the 1920s. He found that the farther a galaxy lay from our own, the faster it was speeding away. In other words, the Universe was expanding.

Hubble's observations were based on analysing the spectra of light from galaxies. A stationary star or galaxy will show a consistent pattern of lines (caused by atoms absorbing particular wavelengths of light in its spectrum).

However, in the case of a moving celestial body, these lines are shifted. The "Doppler Shift" is what you experience when a speeding ambulance rushes towards you; its siren sounds high-pitched, but the pitch drops as it moves

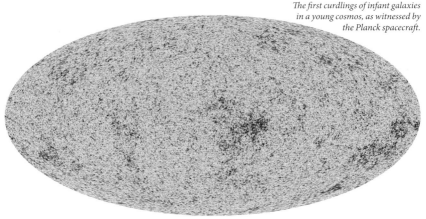

*The first curdlings of infant galaxies in a young cosmos, as witnessed by the Planck spacecraft.*

away. The sound's wavelength is stretched to a lower frequency, and the same is true for light. The spectra of objects moving away are shifted to the red end of the spectrum while those of objects approaching are shifted to the blue. Hubble found that most galaxies had "redshifts," and proved that we live in an expanding cosmos.

The early seconds of the Big Bang were turbulent, to say the least. The temperature was incandescent, the density infernal, but the young Universe had begun to create its subatomic particles. After three minutes, the first elements, hydrogen and helium, had been formed. These would become the building blocks of the stars and galaxies we know today.

Space is still filled with the afterglow of creation, a "microwave background" of radio waves that is the remnant of this birth in fire, cooled down by the relentless expansion to a mere shadow of its former self.

## DARK MATTER

We think of the Universe as a glorious, luminous entity, alight with shining stars, planets and galaxies. But nothing could be further from the truth. In recent years, astronomers have discovered that around 90 percent of our cosmos is invisible, taking the form of mysterious "dark matter."

Although we cannot see it, dark matter makes its presence known through its gravity. It reins in speeding galaxies within galaxy clusters, and it prevents spinning galaxies from flying apart. Astronomers suspect that it is made of some kind of subatomic particles, and researchers around the world are trying to snare some of them in underground laboratories. In theory, these particles were created in the Big Bang, and should still be around today.

# END OF THE UNIVERSE

Although the Universe began in a blaze of glory, its end will be a story of decay. As a result of the Big Bang, the galaxies are now moving apart from one another. Moreover, in the late 1990s, astronomers discovered that the Universe is not just expanding: the rate of expansion is accelerating.

The finding came from observing supernovae in distant galaxies. These are what are termed "standard candles": they are all of the same luminosity and, from their brightness, you can work out how far away they are. The observations show that the most distant galaxies are racing away faster and faster; in the distant future, they will disappear from sight.

What's the cause of this acceleration? The best bet is a new force acting on the Universe: dark energy. Measurements now suggest that a vast proportion of our cosmos is driven by forces that we, as yet, do not understand. Dark matter holds galaxies together; dark energy tears them apart.

We cannot be certain of the ultimate fate of the cosmos. Some scientists think that dark energy may reverse its repulsive effect and everything will collapse into a fireball: the Big Crunch. Others believe that space itself will rip apart and destroy all the matter in it.

But, most likely, the fate of our Universe is simply bleak. All of the galaxies have now been born; in billions of years, they will run out of gas to create new stars. So, the future of the cosmos will be that of an empty space of dying galaxies, many made up of waning black holes.

So, let us celebrate the here and now. The Universe is in its prime. Galaxies are ablaze with nebulae. Magnificent stars are still being born, with a plethora of amazing planets—and with the potential for a whole zoo of alien life.

It couldn't be a better time to be part of our astounding cosmos.

*The Hubble telescope peers into the distant past of the Universe in this deep-field image. Galaxies then were much more distorted than the serene shapes we are familiar with today.*

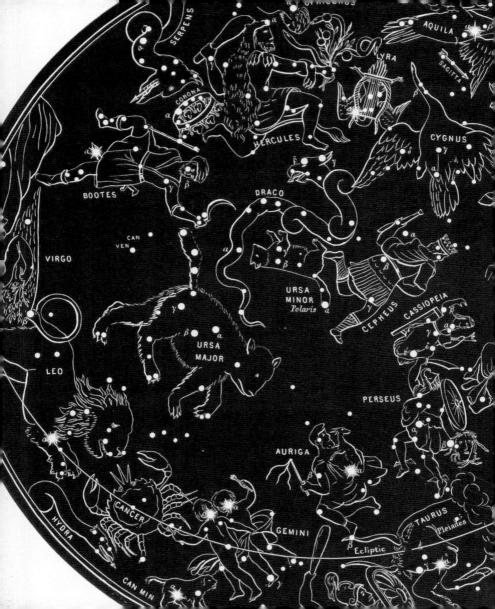

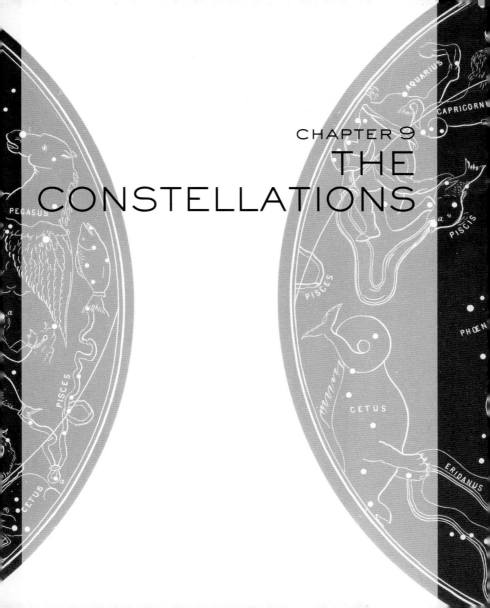

CHAPTER 9
# THE CONSTELLATIONS

# INTRODUCTION

At first sight, the night sky is like a foreign country; it would be easy to get lost in its complex scenery. What is that bright star? What is the nature of that faint fuzzy patch?

Today you can turn to an app on your phone or tablet and it will instantly flag what you are looking at. However, that is like navigating your way on Earth with satnav; you don't learn much about the region you are visiting.

The traditional method, one that has been tried and trusted for millennia, is undoubtedly the best way to get to

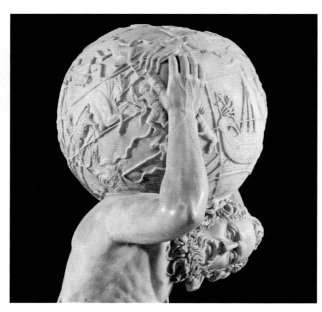

*The Farnese Atlas is a Roman copy of an ancient Greek sculpture, showing the giant Atlas holding a globe of the sky — on which are depicted the 41 most ancient constellation patterns. Atlas's thumb rests on Orion; while Canis Major (the "Great Dog") and the fabled ship Argo lie to the right.*

know the sky intimately. Astronomers have divided up the stars into distinctive patterns: the constellations. Once you are familiar with the brighter constellations, like Orion, Ursa Major or the Southern Cross (Crux), you can use them as signposts to the fainter star patterns.

Turn to the simple sky maps on pages 28–35 to pick out the major star patterns currently visible. Then delve into this chapter to explore each constellation's treasures.

There is an added bonus: each of these celestial shapes has a story to tell. The major constellations, like Leo and Scorpius, have their roots in the old myths of gods and heroes from ancient Mesopotamia and Greece, while more recent constellations include a bird of paradise and even an air pump.

## CHINESE CONSTELLATIONS

The stars in the sky are like a "connect the dots" puzzle book, and in different parts of the world astronomers have connected them in various ways. While the west has 88 constellations, the ancient Chinese drew up 283 smaller star patterns. For example, a distinctive W-shape of stars was known to the Greeks as Cassiopeia, but the Chinese divided it into three constellations, including two paths across the

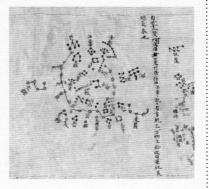

mountains (the Milky Way) and a charioteer drawn by four horses.

# MAPPING THE SKY

Ancient astronomers reveled in the shape of the constellations, to the extent that some actually overlapped. The star Alpheratz, for instance, marks both the head of Princess Andromeda and the navel of the flying horse Pegasus.

In 1930, the International Astronomical Union tidied up the sky. They threw out some superfluous star shapes, including the Sceptre and Hand of Justice, the Office Machine and the poor Cat. They drew immutable borders between the remaining 88 constellations.

On the star charts in this chapter, you will see these constellation boundaries, as well as the "stick figures" defining the constellation shapes. A distinct pattern of stars within a constellation's boundary is known as an asterism, like the Plough within Ursa Major, or the ring of stars within Pisces known as the Circlet.

We have also shown the imaginary lines in the sky that correspond to latitude and longitude on Earth: the distance north or south of the Equator in the sky (celestial latitude) is known as declination, while the distance around

*Astronomers chart the sky in a similar manner to the way that geographers map our world. North and south of the equator in the sky, declination corresponds to latitude on Earth. In the west-east direction, right ascension is the celestial version of longitude; it's measured from the First Point of Aries, where the Sun's path (the Ecliptic) crosses the celestial equator.*

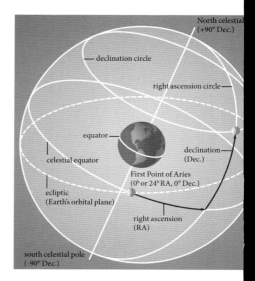

North celestial (+90° Dec.)

declination circle

right ascension circle

equator

declination (Dec.)

celestial equator

First Point of Aries (0ʰ or 24ʰ RA, 0° Dec.)

ecliptic (Earth's orbital plane)

right ascension (RA)

south celestial pole (-90° Dec.)

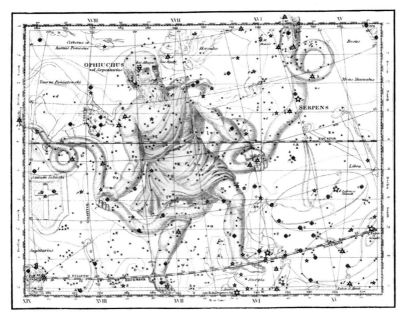

*Less precise but more romantic, the ancients used patterns to map the sky — like Ophiuchus (the serpent-bearer)*

the sky from west to east (celestial longitude) is called right ascension.

There is another important route through the stars: the band of the Zodiac, which is followed by the Moon and the planets. It passes through the traditional 12 "Signs of the Zodiac," from Aries though to Pisces (plus another, Ophiuchus) — although astronomers do not believe in the nonsensical interpretations that astrologers give to these constellations.

We have marked the central line of the Zodiac, the Ecliptic. This is the path we would see the Sun follow during the year, if we could strip away the bright blue sky and view the stars during the day.

*For a definitive list of the constellations, see pages 380-83 — where we also give the genitive versions of the names used in star-naming (for instance, an important variable star in the constellation Cepheus is named Delta Cephei).*

# ANDROMEDA

To the ancient Greeks, this line of three mediocre stars was the centerpiece of vast cosmic drama: Princess Andromeda chained to a rock and about to be gobbled up by a sea monster. Her mother, Cassiopeia, had boasted that she was more beautiful than the sea nymphs, which provoked the sea god to despatch the ravening maritime beast, Cetus. Her father, Cepheus, learned that the only way out was to sacrifice Andromeda to the beast.

As the monstrous Cetus approached, the hero Perseus swooped down from the sky, despatched the monster and married the princess. All of these characters are represented by constellations around Andromeda.

**Almach**, the star at the end of the line, is a beautiful double as seen in a small telescope. The main star is a brilliant yellow supergiant, while its companion is bluish, and the contrasting colors make a lovely sight. Almach is actually a quadruple star, as the companion is a very close triple star.

But the constellation's crowning glory is the **Andromeda Galaxy**. Best seen in a really dark sky away from streetlights when the Moon isn't around, this great star city is the most distant object easily visible to the unaided eye, lying a mind-boggling 2.5 million light years away. To the naked eye, the Andromeda Galaxy is a fuzzy patch of light, bigger than the Moon. You will see it more clearly in binoculars, while a telescope (using a low magnification) shows dark lines across the galaxy and its two large companion galaxies. But you will need to take a long-exposure image of the Andromeda Galaxy to reveal its beautiful spiral structure.

FOR ANTLIA, SEE PAGE 366

FOR APUS, SEE PAGE 336

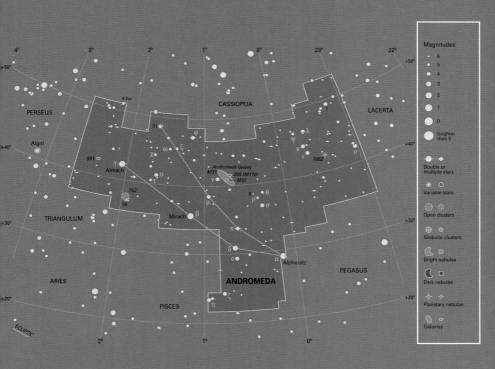

## PROMINENT STARS

| NAME | MAGNITUDE | TYPE | DISTANCE* | LUMINOSITY† | SIZE† |
|------|-----------|------|-----------|-------------|-------|
| Mirach | 2.1 | red giant | 197 | 2,000 | 100 |
| Alpheratz | 2.2 | blue giant | 97 | 240 | 3 |
| Almach a | 2.3 | yellow supergiant | 350 | 2,000 | 80 |
| Almach b | 4.8 | blue-white main sequence | 350 | 70 | 2 |

*light years; †compared to the sun

# AQUARIUS (THE WATER CARRIER)

Aquarius has a pedigree stretching all the way back to the ancient Babylonians, who saw this constellation as the great god Ea pouring water from an overflowing vase. When he got carried away, the tumultuous flow became the devastating floods of the Rivers Tigris and Euphrates.

The surrounding constellations also share a "watery" theme, such as Cetus (the sea monster), Pisces (the fishes) and Capricornus (the sea goat). Ancient astronomers may have associated this entire region of the sky with water because the Sun passed these constellations during the rainy season, from February to March.

By Greek times, Aquarius was merely a man with a water jug, marked by the four central stars of the constellation. However, the aura of divine providence seems to have lingered on, with the Arab names of its brightest stars meaning "the king's lucky star" (**Sadalmelik**) and "luck of the tents" (**Sadachbia**).

**Sadaltager** ("luck of the merchant") is a close pair of white stars. You will need a good backyard telescope to split this duo of celestial diamonds.

Aquarius hosts a glorious sight, as revealed in long-exposure images. The **Helix Nebula** is a planetary nebula, a star in its death throes, and, at 700 light years away, one of the closest to the Earth. It is half the size of the Full Moon in the sky, and visible as a faint celestial ghost in binoculars or through a small telescope.

Fragments of Halley's Comet stream down from Aquarius on or around May 5–6 every year, and are named for one of the stars in the water jug as the **Eta Aquarid** meteor shower.

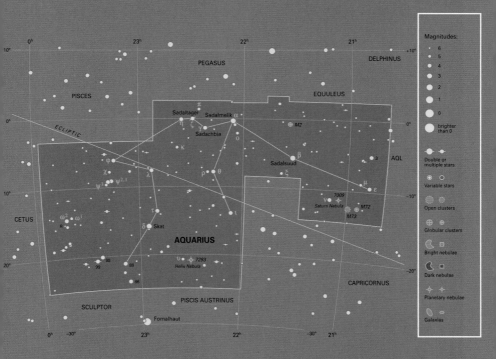

## PROMINENT STARS

| NAME | MAGNITUDE | TYPE | DISTANCE* | LUMINOSITY† | SIZE† |
|------|-----------|------|-----------|-------------|-------|
| Sadalsuud | 2.9 | Yellow supergiant | 540 | 2,300 | 50 |
| Sadalmelik | 3.0 | Red giant | 520 | 3,000 | 80 |

*light years; †compared to the sun

# AQUILA (THE EAGLE), SCUTUM

The brave bird that carried Jupiter's thunderbolts during the war between the gods and the Titans has flown up to the sky as the constellation Aquila. Its other famous mission was to bring Jupiter the most handsome youth in the world, Ganymede.

The constellation is dominated by **Altair**, the 12th-brightest star in the sky; the name means "flying eagle" in Arabic. It is quite easy to identify Altair, as it is flanked by a fainter star to either side. Altair is a young blue-white star only 17 light years away and spins round at a whirlwind rate.

**Eta Aquilae** is one of the brightest Cepheid variable stars, old stars that change in brightness as they swelling and shrinking. Lying 1,400 light years away, Eta Aquilae varies from magnitude 3.5 to 4.4 every seven days.

## SCUTUM (THE SHIELD)

In 1690, Polish astronomer Johannes Hevelius joined up faint stars near Aquila to create the constellation of Scutum Sobiescianum (the Shield of Sobieski), honoring the warrior-king Jan III Sobieski of Poland. Thankfully, the name has now been shortened to Scutum.

The Milky Way appears particularly bright as it passes through Scutum, because this region is clear of the obscuring dust that generally blots out the distant stars.

The constellation's main attraction is the **Wild Duck Cluster**. You can see this compact grouping of 3,000 stars as a fuzzy patch with binoculars, and it is a glorious sight in a small telescope. The cluster forms a triangle, leading the 19th-century British admiral and astronomer William Smyth to say it "somewhat resembles a flight of wild ducks in shape."

FOR ARA, SEE PAGE 332

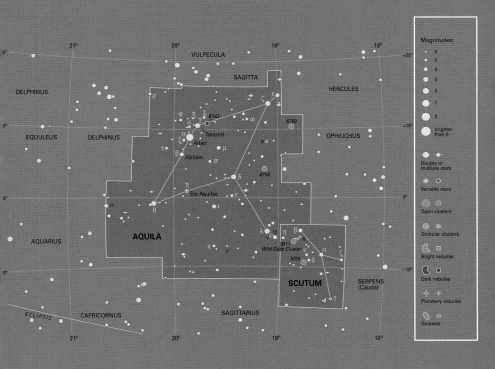

## PROMINENT STARS

| NAME | MAGNITUDE | TYPE | DISTANCE* | LUMINOSITY† | SIZE† |
|------|-----------|------|-----------|-------------|-------|
| Altair | 0.77 | White main sequence | 17 | 11 | 1.8 |
| Tarazed | 2.7 | Orange giant | 395 | 2,500 | 95 |

*light years; †compared to the sun

# ARIES (THE RAM), TRIANGULUM

Aries has only two moderately bright stars (**Hamal** and **Sheratan**), which together with fainter Mesarthim make up the head of the celestial ram. However, it is a very ancient constellation. Around 2,000 years ago, the Sun, on its annual migration from the southern hemisphere to the north, used to cross the celestial equator in Aries. It was a sign that spring was on the way.

In Greek mythology, this beast had an unfortunate ending. It was the "Golden Ram" that rescued the hero Phrixos, only to be sacrificed to the gods by the young man. Phrixos hung his skin in the temple, where it was coveted as the "golden fleece."

Of the three stars marking Aries' head, the faintest, **Mesarthim**, is the most interesting. Lying 165 light years away, this double star has two equally bright white components, easily visible through a small telescope. Mesarthim was the first double star discovered with a telescope, when English scientist Robert Hooke was following a comet, way back in 1664.

## TRIANGULUM (THE TRIANGLE)

Although it is even more obscure than its neighbor, Aries, Triangulum, too, dates back to the earliest times. Its one claim to fame is the **Triangulum Galaxy**, the third-largest member of our Local Group of galaxies (after the Andromeda Galaxy and the Milky Way).

Experienced astronomers can spot this galaxy with the naked eye (under ideally dark and clear skies), even though it lies three million light years away. Binoculars or a small telescope show it as faint fuzzy patch; you will need a large telescope, and really dark conditions, to discern its spiral shape.

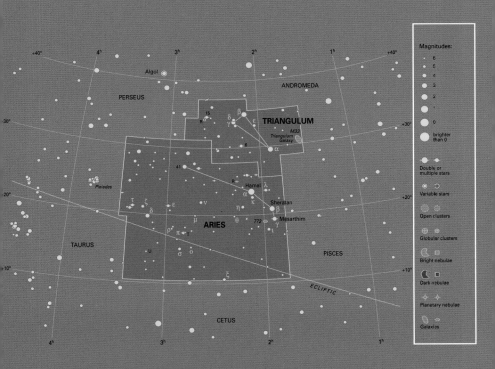

## PROMINENT STARS

| NAME | MAGNITUDE | TYPE | DISTANCE* | LUMINOSITY† | SIZE† |
|------|-----------|------|-----------|-------------|-------|
| Hamal | 2.0 | Orange giant | 66 | 90 | 15 |
| Sheratan | 2.7 | White main sequence | 60 | 23 | 1.8 |

*light years; †compared to the sun

# AURIGA (THE CHARIOTEER), LYNX

Sparkling during the first few months of the year, Auriga is named after the Greek hero Erichthoneus, who invented the four-horse chariot to combat his lameness. The constellation is dominated by **Capella**, the sixth-brightest star in the sky. Its name means "little she goat," but there is nothing diminutive about Capella. It is actually a pair of giant stars orbiting so closely, taking just 104 days per orbit, that we cannot see them separately.

The second-brightest star, **Menkalinan** ("charioteer's shoulder" in Arabic) is an eclipsing binary star, changing in brightness every two days as the two stars pass in front of each other.

Next to Capella you will find a triangle of stars nicknamed "the kids" (the Babylonians saw Auriga as a shepherd's crook): two of these stars are also eclipsing binaries. **Sadatoni** (Zeta Aurigae) is an orange giant star, eclipsed every 972 days by a blue-white partner.

**Almaaz** (Epsilon Aurigae) is one of the weirdest star systems in the sky. Every 27 years, it is eclipsed by what is thought to be a vast dark disc of material swirling around a hidden companion star.

Within the "body" of the Charioteer you will find three very pretty star clusters. Named for their place in the catalogue of nebulous objects drawn up by French astronomer Charles Messier, **M36**, **M37** and **M38** are all visible as fuzzy patches in a pair of binoculars, while a small telescope will display their starry contents.

## LYNX (THE LYNX)

In the 17th century, Polish astronomer Johannes Hevelius created this constellation from faint stars between Auriga and Ursa Major. He said himself that you would need "the eyesight of a lynx" to see it, and the constellation contains little of interest.

FOR "BIG DIPPER," SEE URSA MAJOR, PAGE 362.

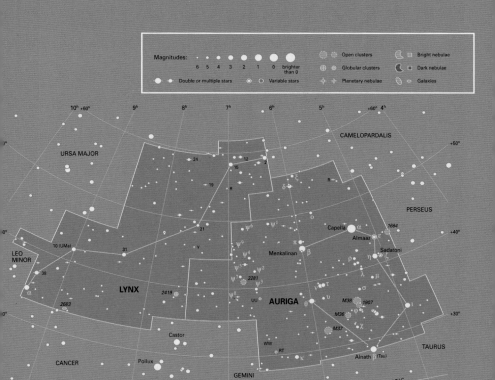

| Magnitudes: | 6 | 5 | 4 | 3 | 2 | 1 | 0 | brighter than 0 | | Open clusters | | Bright nebulae |
| Double or multiple stars | | | | | Variable stars | | | | Globular clusters | | Dark nebulae |
| | | | | | | | | | Planetary nebulae | | Galaxies |

## PROMINENT STARS

| NAME | MAGNITUDE | TYPE | DISTANCE* | LUMINOSITY† | SIZE† |
|------|-----------|------|-----------|-------------|-------|
| Capella | 0.08 | Yellow giant pair | 42 | 9,364 | 12, 9 |
| Menkalinan | 2.7 | White giant pair | 81 | 4,848 | 2.7, 2.6 |

*light years; †compared to the sun

# BOÖTES (THE HERDSMAN), CORONA BOREALIS

A kite-shaped constellation, mentioned in Homer's *Odyssey*, Boötes is the celestial herdsman in charge of the stars in the northern part of the sky.

The name of **Arcturus**, the brightest star, actually means "bear-driver," because it apparently "drives" the two bears (Ursa Major and Ursa Minor) as the Earth rotates. Arcturus is the fourth-brightest star in the sky. It was an important navigational beacon for ancient Polynesian sailors crossing the Pacific, as it passes directly over Hawaii.

Through a good telescope, **Izar** (the belt) appears as a double star, one star yellow and the other blue. It is also called Pulcherrima, meaning "most beautiful."

The rather fainter **Alkalurops** (the herdsman's crook) is a lovely triple star. Through binoculars it is double, while a telescope shows the fainter star is itself a close double.

A brilliant display of shooting stars rains down from Boötes on January 3–4. Confusingly, they are called the **Quadrantids**, as this region was once known as Quadrans Muralis (the Mural Quadrant, an astronomical instrument).

## CORONA BOREALIS (THE NORTHERN CROWN)

Small but perfectly formed, **Corona Borealis** depicts the crown that Bacchus gave Ariadne on their wedding. The heavenly tiara is studded with an ultimate jewel, the blue-white star **Gemma**. There are also two bizarre variable stars.

**R Coronae Borealis** normally hovers around the limits of naked-eye visibility, but it occasionally disappears behind veils of sooty clouds. **T Coronae Borealis** behaves in the opposite way. Usually skulking beyond the range of binoculars, this star has sudden outbursts when it surges to rival Gemma.

FOR CAELUM, SEE PAGE 352

FOR CAMELOPARDALIS, SEE PAGE 304

## PROMINENT STARS

| NAME | MAGNITUDE | TYPE | DISTANCE* | LUMINOSITY† | SIZE† |
|------|-----------|------|-----------|-------------|-------|
| Arcturus | -0.04 | Orange giant | 37 | 170 | 26 |
| Gemma | 2.2 | White main sequence | 75 | 75 | 3 |

*light years; †compared to the sun

# CANCER (THE CRAB)

Cancer is the faintest of the constellations of the Zodiac, the band through which the Sun, Moon and planets appear to move. City lights completely drown out the constellation, but under dark skies you will find the celestial crab between the Sickle of Leo and the twin stars Castor and Pollux in Gemini.

According to legend, Cancer tried to distract Hercules by nipping his ankle during the hero's altercation with the multi-headed monster Hydra, one of his "12 labors." When Hercules crushed the crustacean under his foot, the goddess Juno (Hercules' bitter enemy) elevated Cancer to the heavens.

Its faint stars are interesting only for their names. The central star holds the record for the longest name: **Arkushanangarushashutu** (Babylonian for "the south-east star of the Crab"), although it is usually known as **Asellus Australis** (the southern ass). Along with **Asellus Borealis** (the northern ass), it flanks Cancer's crowning glory, the Manger.

You can easily spot the Manger (officially known by its Latin name, **Praesepe** ) with the naked eye, on a clear, dark night, when it appears as a faint, fuzzy patch of light. Peering through his pioneering telescope, Galileo found Praesepe was a "mass of more than 40 stars." Those swarming stars have led to its more common modern name, the **Beehive Cluster**. The Beehive is so large that it's best to view it through binoculars or a telescope at very low magnification.

You will certainly need a telescope to see the stars in Cancer's other star cluster, **M67**, which lies 2,700 light years away.

FOR CANES VENATICI, SEE PAGE 362

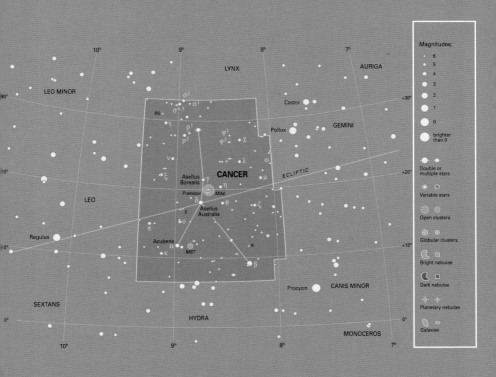

# CANIS MAJOR (THE GREAT DOG)

Crowned by brilliant Sirius, Canis Major is the larger of Orion's two hunting dogs. He is chasing Lepus (the Hare), but his main target is Orion's chief quarry, Taurus (the Bull). The Indians regarded this constellation and Canis Minor as the "watchdogs of the Milky Way," which runs between the two star patterns.

**Sirius** is the brightest star in the sky. It is twice as massive and 25 times more luminous than our Sun. From the earliest times, Sirius has been known as the Dog Star, giving its name to the entire constellation. Roman farmers sacrificed dogs when the Sun approached the star in May. The debilitating "dog days of summer" were ascribed to Sirius' irradiance adding to the Sun's heat. To the ancient Egyptians, Sirius was crucial, because the first time each year that it could be seen rising just before the Sun foretold the annual flooding of the Nile.

Technically, astronomers call this star **Sirius A**; because it has a small companion star: a white dwarf almost as heavy as the Sun, but no bigger than the Earth. To spot this star (**Sirius B**, or "the Pup"), you will need quite a powerful backyard telescope (bigger than 200 mm).

Near Sirius lies the beautiful star cluster **M41**. This loose agglomeration of over a hundred young stars 2,300 light years away is easily visible through binoculars, and even to the unaided eye. It was rumored that the Greek philosopher Aristotle, in 325 BC, called it "a cloudy spot," which would be the earliest surviving description of a deep-sky object.

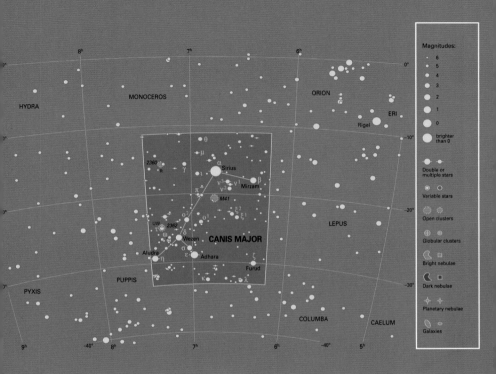

## PROMINENT STARS

| NAME | MAGNITUDE | TYPE | DISTANCE* | LUMINOSITY† | SIZE† |
|------|-----------|------|-----------|-------------|-------|
| Sirius A | -1.47 | White main sequence | 8.58 | 25 | 1.7 |
| Sirius B | 8.3 | White dwarf | 8.58 | 0.026 | 0.01 |
| Adhara | 1.5 | Blue-white supergiant | 430 | 39,000 | 14 |
| Wezen | 1.8 | Yellow-white supergiant | 1,800 | 80,000 | 200 |
| Mirzam | 2.0 | Blue-white giant | 500 | 27,000 | 10 |

*light years; †compared to the sun

# CANIS MINOR (THE LITTLE DOG), MONOCEROS

In every way junior to its sibling, Canis Major, the Little Dog has one prominent star, and that's about it. The Greeks named this star **Procyon**, meaning "before the dog" because its rising foretold the appearance of Sirius.

To the ancient Babylonians, this constellation was a rooster, but the Greeks turned these stars into the second of Orion's two hunting dogs. Another myth relates that it is Maera, the faithful hound of Icarius, the original wine-maker. When he invited some shepherds for the first-ever drinks party, they became intoxicated and thought Icarius had poisoned them. The shepherds murdered their generous host and buried his body. But Maera revealed her master's grave, the shepherds were hanged and the Little Dog earned her place among the stars.

Like the Dog Star, Procyon has a white dwarf companion, but **Procyon B** is even fainter than Sirius's "Pup," needing a large telescope to be seen at all.

## MONOCEROS (THE UNICORN)

In 1612, Dutch cartographer and theologian Petrus Plancius created a constellation to honor the unicorn, which is mentioned in the Bible as a symbol of strength. **Monoceros** has no bright stars, but binoculars or a telescope will reveal some interesting sights.

**M50** is a heart-shaped star cluster, visible with binoculars; a telescope reveals a scattering of celestial sapphires, set with one ruby. Another triangular group of scintillating stars looks, though a telescope, like a set of Yuletide decorations, so it is named the **Christmas Tree Cluster**.

The glory of the constellation is the aptly named **Rosette Nebula**, but its spectacular red petals are only visible in long-exposure images. Through the telescope, you will just see the central cluster of stars.

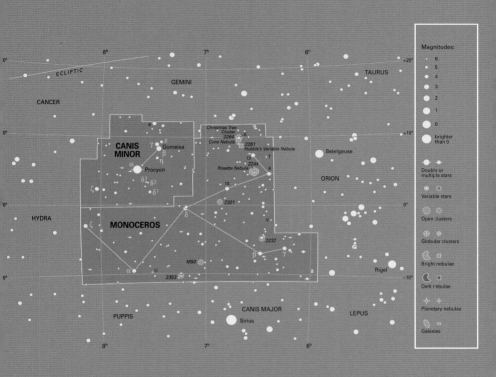

## PROMINENT STARS

| NAME | MAGNITUDE | TYPE | DISTANCE* | LUMINOSITY† | SIZE† |
|------|-----------|------|-----------|-------------|-------|
| Procyon A | 0.34 | Yellow-white main sequence | 11.4 | 7 | 2 |
| Procyon B | 10.7 | White dwarf | 11.4 | 0.0005 | 0.01 |

*light years; †compared to the sun

# CAPRICORNUS (THE SEA GOAT)

Despite its dim outline, Capricornus was one of humanity's first constellations. It is among a group of wet and watery star patterns that swim in the celestial sea near Pegasus, the winged horse. Capricornus had a special significance for the ancient people of the Middle East, 2,500 years ago who, rather bizarrely, saw this large and obscure triangle of stars as a goat with a fish's tail. At that time, the Sun nestled amongst the stars of Capricornus at the Winter Solstice, showing the astronomers that the year was about to turn around; the longest nights were at an end, and life-giving spring was on the way. It has also given its name to the Tropic of Capricorn, the imaginary line on the Earth marking the southern limit of the zone where the Sun can appear overhead.

**Algedi** is the most interesting star in Capricornus, lying at one corner of the triangle. Even with the unaided eye, you can see that this faint star is double. It is just a chance alignment, though, as the brighter star lies 109 light years away, and the fainter is 690 light years from us. A telescope will show you that each star is in fact a genuine double.

By coincidence, the next-door star **Dabih** is also a double. The main member of this duo is a yellow star, while binoculars or a small telescope will reveal a fainter blue companion.

A telescope is also essential for the next beast in Capricornus, the globular cluster **M30**. Lying about 26,000 light years away, this rather ragged ball of thousands of stars was probably among the first objects to form in our Galaxy. And it is very pretty — a lovely target for astrophotographers.

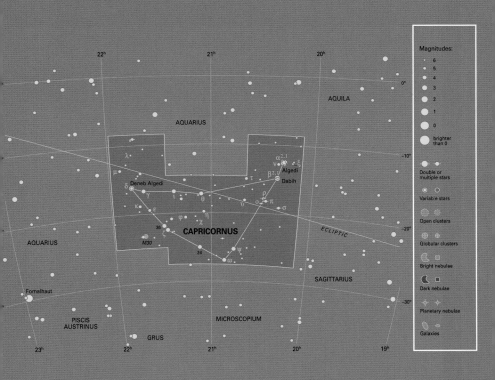

## PROMINENT STARS

| NAME | MAGNITUDE | TYPE | DISTANCE* | LUMINOSITY† | SIZE† |
|------|-----------|------|-----------|-------------|-------|
| Deneb Algedi | 2.8 | White giant | 38 | 8 | 2 |

*light years; †compared to the sun

# CARINA (THE KEEL)

Sailing along the band of the Milky Way, Carina is rich in beautiful sky sights. To the Greeks, it was the keel of the great ship *Argo*, a vast constellation dismembered in the 18th century (see Vela, page 366).

The jewel in Carina's crown is **Canopus**, the second-brightest star in the night sky. Marking the *Argo*'s rudder, this star is named after a legendary Greek navigator. For most of the past four million years, Canopus has been our brightest star; Sirius has only taken that title temporarily as it has zoomed closely past the Solar System.

The beautiful star cluster of the **Southern Pleiades** (IC 2602) is a close rival to the original Pleiades in Taurus. It is a fantastic sight through binoculars. Sweep along the Milky Way to spot other lovely star clusters.

The brightest member of the Southern Pleiades forms one corner of the **Diamond Cross**, a perfect lozenge that also includes **Miaplacidus** (Carina's second most prominent star).

Speaking of jewels, the **Carina Nebula** is the most spectacular nebula in the sky. It is easily visible to the naked eye as a glowing cloud in the Milky Way, and an incredible sight through binoculars or a telescope.

At the nebula's heart lurks a celestial monster, **Eta Carinae**. This star is a hundred times heavier than the Sun, and in its death throes. Currently, Eta Carinae is just visible to the unaided eye, but in the 1840s it brightened until it almost rivalled Sirius. The star is now swathed in dust from the eruption; the likelihood is that it will explode as a supernova within a few thousand years, to blaze in our skies 20 times brighter than Venus.

**Avior** and **Aspidiske** — with **Delta Velorum** and **Markeb** in Vela (page 366) — make up the **False Cross**: it's significantly bigger than the real Southern Cross (page 312).

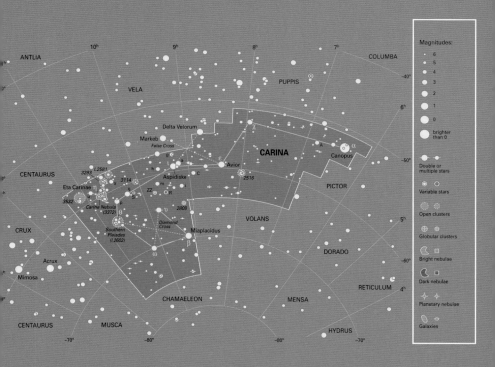

## PROMINENT STARS

| NAME | MAGNITUDE | TYPE | DISTANCE* | LUMINOSITY† | SIZE† |
|------|-----------|------|-----------|-------------|-------|
| Canopus | -0.72 | Yellow-white supergiant | 310 | 15,000 | 70 |
| Miaplacidus | 1.7 | White giant | 113 | 290 | 7 |
| Avior | 1.9 | Orange giant | 610 | 5,000 | 20 |

*light years; †compared to the sun

# CASSIOPEIA, CAMELOPARDALIS

To the ancients, this W-shaped star pattern represented Queen Cassiopeia. The wife of King Cepheus, she misguidedly boasted that her daughter Andromeda was more beautiful than the sea nymphs. The sea god sent a ravaging monster (Cetus) to eat the young people of the country. It could only be appeased by the sacrifice of Andromeda — but she was rescued by the hero Perseus. The main characters are now immortalized in the heavens.

This constellation has seen its share of real cosmic drama. In 1572, Danish astronomer Tycho Brahe was amazed to observe a supernova that rivalled Venus in brightness. A dimmer supernova erupted in around 1660, and its expanding gases form the most prominent celestial radio source in the sky, Cassiopeia A.

The Chinese saw Cassiopeia as three star groups, including a chariot and a mountain path. Unusually, the constellation's central star is universally known today by its Chinese name, **Tsih** (the whip). Some 55,000 times brighter than the Sun, it spins around at breakneck pace, flinging out streams of gas.

With binoculars or a small telescope, seek out the star clusters **M52** and **M103**.

### CAMELOPARDALIS (THE GIRAFFE)

Around 1612, the religious Dutch sky-mapper Petrus Plancius drew the outline of a giraffe in these faint stars, its name combining the Latin for the long-necked camel and spotted leopard. He intended to immortalize the biblical camel that Rebecca rode to meet her future husband, Isaac, but somehow her celestial transport became an even more uncomfortable giraffe.

The only object of interest is **NGC 2403**, a lovely spiral galaxy in a small telescope. The Voyager 1 spacecraft is heading into interstellar space in this direction.

## PROMINENT STARS

| NAME | MAGNITUDE | TYPE | DISTANCE* | LUMINOSITY† | SIZE† |
|---|---|---|---|---|---|
| Schedir | 2.2 | Orange giant | 230 | 680 | 42 |
| Caph | 2.3 | Yellow-white giant | 55 | 27 | 4 |

*light years; †compared to the sun

# CENTAURUS (THE CENTAUR)

To the ancient Greeks, it was a mythical creature, with the body of a horse, and a man's torso and head. Originally, the stars of the Southern Cross (Crux) were the centaur's feet, making the ancient constellation even more splendid.

**Alpha Centauri** is the third-brightest star in the sky (magnitude -0.27). Even a small telescope shows that it is a close-set pair of brilliant stars. These twins are the closest stars visible to the naked eye, at 4.37 light years; their dim companion, fittingly named **Proxima Centauri**, is our nearest stellar neighbor, at 4.24 light years away.

Together with the constellation's second star, **Hadar** (sometimes known as **Agena**), Alpha Centauri is one of "The Pointers" that directs you towards the Southern Cross.

**Omega Centauri** appears as a hazy star to the unaided eye. In reality, it is the most massive star cluster in our Galaxy, swarming with millions of stars and lying 16,000 light years away. It is a glorious sight through a telescope of any size.

Near to Omega in the sky, but almost a thousand times further away, lies **Centaurus A**. The fifth-brightest galaxy in our skies, this giant elliptical contains a trillion stars. It can be seen through binoculars, but a small telescope reveals a dark band: the remains of a small dusty galaxy being consumed by the giant. Gas falling towards a massive black hole at the heart of Centaurus A releases huge amounts of energy and makes this galaxy, despite its vast distance, one of the strongest radio sources detectable from Earth.

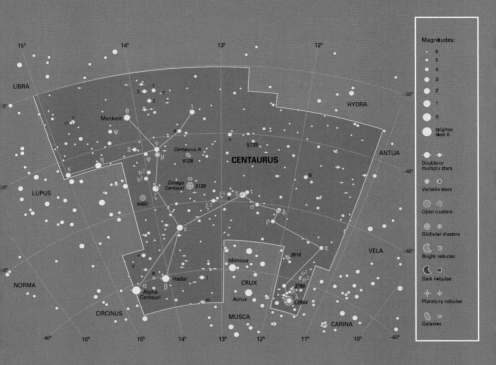

| Magnitudes: |
| · 6 |
| · 5 |
| • 4 |
| ● 3 |
| ● 2 |
| ● 1 |
| ● 0 |
| ● brighter than 0 |

Double or multiple stars

Variable stars

Open clusters

Globular clusters

Bright nebulae

Dark nebulae

Planetary nebulae

Galaxies

## PROMINENT STARS

| NAME | MAGNITUDE | TYPE | DISTANCE* | LUMINOSITY† | SIZE† |
|---|---|---|---|---|---|
| Alpha Centauri A | -0.01 | Yellow main sequence | 4.4 | 1.5 | 1.2 |
| Alpha Centauri B | 1.33 | Orange main sequence | 4.4 | 0.5 | 0.9 |
| Hadar | 0.60 | Blue-white giants (pair) | 350 | 20,000 | 10 |

*light years; †compared to the sun

# CEPHEUS

Shaped like a child's drawing of a house, this star-pattern is named after the King of Aethiopia of Greek mythology, married to the far more magnificent next-door constellation, Cassiopeia. Both in legend and visually, his wife is far more exciting.

As a constellation, Cepheus is faint and somewhat boring, save for a trio of fascinating stars. **Alfirk** is a lovely double star; the companion is visible through a small telescope.

The aptly known **Garnet Star**, named by the 18th-century British astronomer William Herschel after its ruddy hue, varies from magnitude 3.4 to 5.1 in the space of two to three years. It is the reddest of the naked-eye stars; through binoculars, it does indeed look like a sparkling gem. The Garnet Star is faint in our skies only because of its distance, an immense 6,000 light years. In fact, it is one of the brightest and biggest inhabitants of our Galaxy, half a million times more luminous than our Sun and so large that if you put it in the Solar System its surface would lie farther out than Jupiter's orbit.

And Cepheus is home to the iconic variable star — **Delta Cephei**. This star changes in brightness (from magnitude 3.5 to 4.4) as it grows and shrinks over a period of 5 days and 9 hours. (Through a small telescope, you can also see it has fainter companion.) Astronomers have discovered that variable stars of this type (named Cepheids) have an unusual property: the time they take to wax and wane is related to their intrinsic luminosity, which allows them to be used as pulsating stellar beacons to measure cosmic distances.

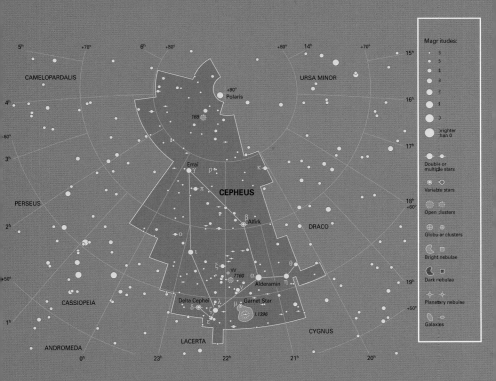

## PROMINENT STARS

| NAME | MAGNITUDE | TYPE | DISTANCE* | LUMINOSITY† | SIZE† |
|------|-----------|------|-----------|-------------|-------|
| Alderamin | 2.5 | White main sequence | 49 | 17 | 2.3 |

*light years; †compared to the sun

# CETUS (THE SEA MONSTER; THE WHALE), SCULPTOR

In the Andromeda legend, Perseus slew this unfortunate sea monster before it could consume the chained-up princess. Cetus swims on the edge of the watery region of the sky that also features the river Eridanus, Pisces (the Fishes) and Aquarius (the Water Carrier).

Cetus has only one star of real note. **Mira** ("The Wonderful") was discovered in 1596 by German astronomer David Fabricius. Over an 11-month period, it varies in brightness alarmingly, from magnitude three to 10. In 1779, William Herschel even saw it as bright as Arcturus. Mira is a distended and unstable red giant. At its biggest and brightest, Mira is 1,500 times brighter than the Sun and 400 times wider.

## SCULPTOR (THE SCULPTOR)

In the 1750s, French astronomer Nicolas Louis de Lacaille "connected the dots" in this barren region of sky, to create the Sculptor's Studio. A century later, British astronomer John Herschel (son of the more famous William) shortened the name to Sculptor.

The faint stars of Sculptor hold little interest, but there is one lovely galaxy, **NGC 253**, discovered by Caroline Herschel (William's sister). The seventh-brightest galaxy in the sky, it is easily visible with binoculars. Through a moderate telescope, you will be treated to a gorgeous spiral galaxy viewed at a steep angle and see why it is nicknamed the Silver Coin Galaxy. **NGC 55** is a slightly fainter edge-on spiral.

FOR CHAMAELEON, SEE PAGE 336

FOR CIRCINUS, SEE PAGE 332

FOR COLUMBA, SEE PAGE 352

FOR COMA BERENICES, SEE PAGE 368

FOR CORONA AUSTRALIS, SEE PAGE 354

FOR CORONA BOREALIS, SEE PAGE 292

FOR CORVUS, SEE PAGE 326

FOR CRATER, SEE PAGE 326

## PROMINENT STARS

| NAME | MAGNITUDE | TYPE | DISTANCE* | LUMINOSITY† | SIZE† |
|------|-----------|------|-----------|-------------|-------|
| Diphda | 2.0 | Orange-yellow giant | 96 | 140 | 17 |

*light years; †compared to the sun

# CRUX (THE SOUTHERN CROSS), MUSCA

It is the emblem of the southern hemisphere, emblazoned on the flags of Australia, New Zealand and Brazil. Oddly, the ancient Greeks regarded this prominent star pattern merely as the feet of Centaurus (the Centaur). In the 16th century, European navigators found it a useful navigational aid, and in 1589 Dutch cartographer and theologian Petrus Plancius first depicted these stars as the Southern Cross (although it more closely resembles a lop-sided kite).

Crux is the smallest constellation in the sky, but one of the richest. The brilliant blue-white stars **Acrux** and **Mimosa** form a splendid contrast to red giant **Gacrux**. Through a telescope, you will see that Acrux has a close companion star.

The most prominent feature of Crux, however, is not the stars but what lies between them. On a dark night, you cannot miss the yawning black gap in the glowing band of the Milky Way. Aboriginal astronomers saw what is now aptly known as the **Coalsack** as the head of a dark emu in the sky. It is the silhouette of a pair of dense dust clouds, lying 610 and 790 light years away.

The **Jewel Box** is a cluster of stars that well deserves its name. In the words of Victorian astronomer John Herschel, a small telescope reveals "a casket of variously colored precious stones." Studded with red and blue-white giant and supergiant stars, it is one of the youngest star clusters known, with an age of "only" 10 million years.

### MUSCA (THE FLY)

Created by Petrus Plancius (see Crux, above) in 1597, Musca is the only constellation that depicts an insect. That is its sole claim to fame.

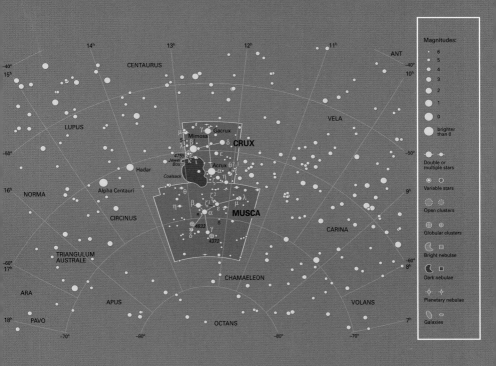

## PROMINENT STARS

| NAME | MAGNITUDE | TYPE | DISTANCE* | LUMINOSITY† | SIZE† |
|------|-----------|------|-----------|-------------|-------|
| Acrux | 0.77 | Blue-white main sequence | 320 | 25,000 | 10 |
| Mimosa | 1.25 | Blue-white main sequence | 280 | 34,000 | 8 |
| Gacrux | 1.6 | Red giant | 89 | 1,500 | 84 |

*light years; †compared to the sun

# CYGNUS (THE SWAN), LACERTA

The glorious celestial swan flies along the Milky Way with outspread wings and outstretched neck, its head marked by Albireo and its tail by brilliant Deneb. In Greek myth, Zeus disguised himself as a swan to seduce Leda, the wife of King Tyndareus of Sparta. The unfortunate woman gave birth to a mix of immortal children (fathered by Zeus) and her husband's mortal offspring. The brood included Helen of Troy, and Pollux and Castor, the heavenly twins in the constellation of Gemini.

**Deneb** (meaning "tail") is a celestial beacon. It is the farthest away of the top 20 brightest stars, so remote that astronomers are unsure of its distance. Estimates range from 1,500 to 2,600 light years, meaning that it shines anything from 50,000 to 200,000 times brighter than the Sun!

The swan's head is marked by probably the most beautiful double star in the sky: **Albireo**. You can just split the pair with good binoculars, and the sight of the gold and blue stars is sensational through a small telescope.

The **North America Nebula** is a cloud of glowing gas, bigger than the Full Moon, but so faint that you will need to use binoculars on a really dark night. While you have them to hand, sweep along the band of the Milky Way, as Cygnus is riddled with star clusters and nebulae. Near Albireo, the luminous band is split by a vast ribbon of dark interstellar dust, the **Cygnus Rift**.

## LACERTA (THE LIZARD)

This dim constellation, created by Polish astronomer Johannes Hevelius in 1697, contains little of interest.

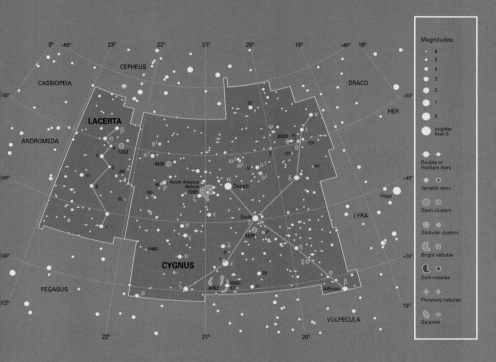

## PROMINENT STARS

| NAME | MAGNITUDE | TYPE | DISTANCE* | LUMINOSITY† | SIZE† |
|------|-----------|------|-----------|-------------|-------|
| Deneb | 1.25 | White supergiant | c. 2,000 | c. 100,000 | c. 150 |
| Albireo A | 3.1 | Yellow giant | 400 | 1,200 | 70 |
| Albireo B | 5.1 | Blue-white main sequence | 400 | 230 | 3 |

*light years; †compared to the sun

# DELPHINUS (THE DOLPHIN), SAGITTA, VULPECULA

It may be small, but it is perfectly formed. **Delphinus**, the celestial dolphin, is outlined by a lopsided rectangle of four stars, with an extra star forming his tail.

This constellation immortalizes humanity's long relationship with the most intelligent marine life on our planet. In one myth, the dolphin acted as go-between when the sea god Poseidon (Neptune) was courting his wife, the sea nymph Amphitrite; in another, the dolphin rescued the musician Arion when he was thrown overboard by sailors intent on stealing his wealth.

The strange names of the two brightest stars, **Sualocin** and **Rotanev**, represent a bit of self-promotion by the 19th-century Italian astronomer, Niccolo Cacciatore. In Latin, his name becomes Nicolaus Venator. Try spelling this backwards.

**Gamma Delphini** is a lovely double star when you observe it with a reasonable telescope.

### SAGITTA (THE ARROW)
The shape of this tiny constellation clearly reflects its name. In Greek myth, Sagitta was the arrow Hercules shot at the eagle (the neighboring constellation Aquila) that was pecking at the liver of the giant Prometheus. The only interesting object, astronomically speaking, is the star cluster **M71**, which is 12,000 light years away and 10 billion years old.

### VULPECULA (THE LITTLE FOX)
Created by Polish astronomer Johannes Hevelius in the 17th century, this faint constellation was originally the "fox and goose," but the bird has subsequently escaped. It shot to fame in 1967, as the lair of the first known pulsar, a rapidly spinning star corpse detected by its regular radio pulse.

With binoculars, you can pick out the group of stars known, for its distinctive shape, as **The Coathanger**. But the constellation's glory is the **Dumbbell**

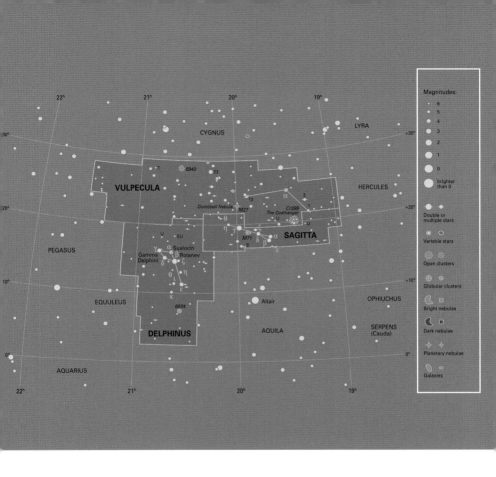

**Nebula**, visible with binoculars and a great sight in a small telescope. The Dumbbell is a planetary nebula, a shell of gas puffed off by a dying star 10,000 years ago.

# DORADO (THE GOLDFISH), MENSA, PICTOR, RETICULUM

You wouldn't keep Dorado at home in a goldfish bowl: this constellation represents the gold-flanked dolphinfish (mahi-mahi), which can grow as big as a human. It was created by Dutch sky-mapper Petrus Plancius in 1597.

This otherwise obscure star pattern boasts one of the greatest of all sky sights: the **Large Magellanic Cloud**. At a distance of 160,000 light years, it is our nearest major galaxy and appears to the naked eye as a prominent glowing patch, like a detached portion of the Milky Way.

You will spot the brightest nebulae and star clusters with binoculars; if you use a telescope, it is easy to get lost in a profusion of detail. The Large Magellanic Cloud's greatest glory is the **Tarantula Nebula**. If you put this brilliant star factory as close as the Orion Nebula, it would appear as bright as the Moon. A massive star exploded here, as **Supernova 1987A**: the first supernova to be visible to the unaided eye in almost 400 years.

## MENSA (TABLE MOUNTAIN)

French astronomer Nicolas Louis de Lacaille drew this shape to commemorate Table Mountain, in South Africa, from where he made his observations in the 1750s. The milky glow of the Large Magellanic Cloud, spilling over from Dorado, represents the cloudy "table cloth" of cloud that often drapes the flat mountain.

## PICTOR (THE PAINTER'S EASEL)

One of Lacaille's inventions (see Mensa, above), Pictor contains only one notable object. **Beta Pictoris** was the first star discovered to have a surrounding disc of dust and gas: a planetary system in formation. Astronomers later found a newly born planet in the disc.

## RETICULUM (THE EYEPIECE GRATICULE)

Originally introduced into the sky as Rhombus, referring to its diamond shape,

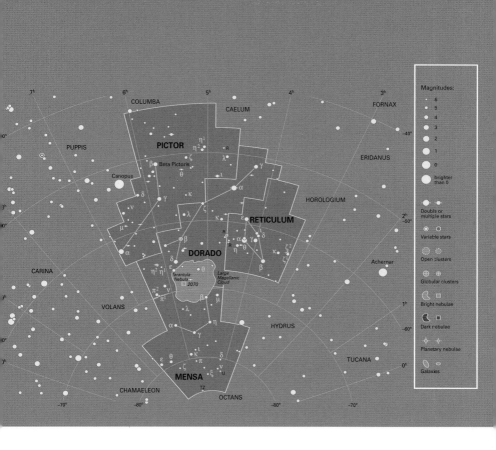

by Isaac Habrecht II of Strasbourg (now in France) in 1621, this constellation was renamed by Lacaille (see Mensa, above), to immortalize the graticule in the eyepiece of his telescope, which he used to measure star positions.

FOR DRACO, SEE PAGE 364

FOR EQUULEUS, SEE PAGE 344

# ERIDANUS (THE RIVER), FORNAX, HOROLOGIUM

A long, winding constellation, the celestial river rises in the star **Cursa** ("footstool") next to brilliant Rigel in Orion, and flows down south below the horizon as seen by ancient astronomers. Eridanus might be named after the ancient city of Eridu, at the mouth of the Euphrates. To Greek astronomers, it was the River Po, in Italy, where the reckless youth Phaethon fell to Earth after losing control of the Sun chariot.

**Acamar**, the southernmost star in Eridanus visible from Greece, was the river's original mouth. A small telescope shows that it is a close pair of white stars.

When European explorers traveled southwards in the Age of Discovery, they spotted brighter **Achernar** south of Acamar, and extended the river so that Achernar became its mouth. (Both names come from the Arabic for "mouth of the river.") Achernar is the 11th-brightest star, and it appears even more prominent because no other bright stars are nearby.

It spins so fast that it is one of the most flattened stars known — 50 percent wider across its equator than pole-to-pole.

**Epsilon Eridani** is among the closest stars, just 10.52 light years away. It's surrounded by a dusty disc, a planetary system in formation, probably containing a planet 50 percent more massive than Jupiter.

With a telescope, find **NGC 1291**, a spiral galaxy surrounded by a ring of stars.

### FORNAX (THE FURNACE)

This constellation was introduced by French astronomer Nicolas Louis de Lacaille in 1754 to immortalize the chemical furnace. The brightest star, **Dalim**, has a close companion.

### HOROLOGIUM (THE CLOCK)

The elongated shape of Horologium depicts a pendulum clock. Another obscure constellation created by Lacaille (see Fornax, above), Horologium contains nothing of interest in a small telescope.

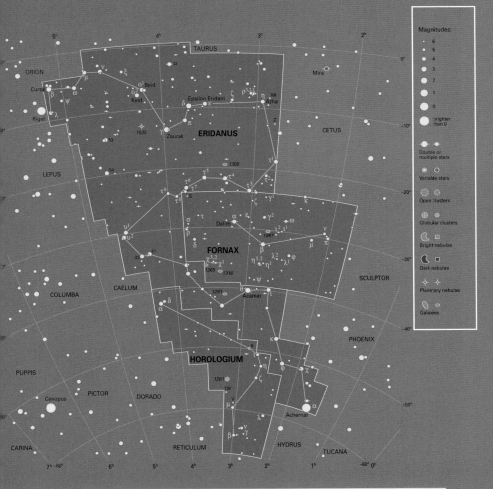

| Magnitudes: |
| 6 |
| 5 |
| 4 |
| 3 |
| 2 |
| 1 |
| 0 |
| brighter than 0 |

Double or multiple stars

Variable stars

Open clusters

Globular clusters

Bright nebulae

Dark nebulae

Planetary nebulae

Galaxies

## PROMINENT STARS

| NAME | MAGNITUDE | TYPE | DISTANCE* | LUMINOSITY† | SIZE† |
|------|-----------|------|-----------|-------------|-------|
| Achernar | 0.50 | Blue-white main sequence | 139 | 3,100 | 7 |

*light years; †compared to the sun

# GEMINI (THE TWINS)

You cannot mistake Gemini. The constellation is crowned by the bright stars Castor and Pollux, representing the heads of a pair of twins with their bodies running in parallel lines of stars. In legend, Castor and Pollux were conceived by the princess Leda on the night she married the King of Sparta, the father of mortal Castor. But Zeus also invaded the marital suite, disguised as a swan, and fathered the immortal Pollux. The pair were so devoted that Zeus placed them together for eternity, among the stars.

**Castor** is an amazing star; it is actually a family of six suns. Through a small telescope, you can see that Castor is a double. Each of these stars has a fainter companion (although you need special equipment to detect this). There is another outlying star, too, visible through a telescope, and this is also double. How ironic that one of the mythological heavenly twins is a triplet of twin stars!

Slightly brighter **Pollux** is an orange giant, its color contrasting nicely with white Castor. Pollux is rare among giant stars in possessing a planet — a massive world bigger than Jupiter.

The star cluster **M35** lies 2,800 light years away, but it is still bright enough to catch with the naked eye. As big as the Full Moon, M35 is a fine sight in binoculars, and a true celestial beauty when viewed through a small telescope.

The **Geminid** meteors rain down from this constellation every year in mid-December. The bright and abundant shooting stars are debris shed by the asteroid Phaethon.

FOR GRUS, SEE PAGE 350

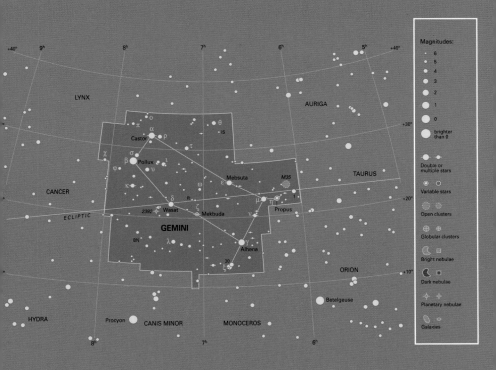

## PROMINENT STARS

| NAME | MAGNITUDE | TYPE | DISTANCE* | LUMINOSITY† | SIZE† |
|------|-----------|------|-----------|-------------|-------|
| Pollux | 1.15 | Orange giant | 34 | 43 | 9 |
| Castor A | 1.9 | White main sequence | 51 | 37 | 2.3 |
| Castor B | 3.0 | White main sequence | 51 | 13 | 1.6 |
| Alhena | 1.9 | White giant | 110 | 120 | 3 |

*light years; †compared to the sun

# HERCULES

For one of antiquity's superheroes, the celestial version of Hercules is pretty feeble. While the hunter Orion, a minor character in Greek myths, has a constellation that is all strutting masculinity, Hercules is but a poor reflection, and upside-down, too. The faint stars here do not do justice to the original superman, famous for his 12 heroic labors.

In fact, this constellation has its roots much earlier than the Greeks. It was originally known as "the Kneeler," with his knees on Draco (the Dragon). The Greek poet Panyassis identified it as Hercules, grappling with a serpent that guarded a grove of sacred golden apples.

Dig a little deeper, however, and you will find a fascinating constellation. Outside the rectangular main "body" of the hero, to the south, you will see **Rasalgethi**, Hercules's head. At around 400 times the Sun's girth, this red supergiant is one of the biggest stars known. Shrinking and billowing in its death throes, Rasalgethi varies in brightness over a period of four months. With a small telescope, you can make out a fainter companion star.

Hercules boasts one of the most spectacular sights in the northern night sky. A small fuzzy patch to the naked eye, **M13** is in reality a giant globular cluster made of almost a million stars, some of the oldest inhabitants of our Galaxy. In the hope that there might be a profusion of inhabited planets in M13, this cluster was the target, in 1974, of the first deliberate radio message from Earth to possible alien civilizations. M13 is sensational through any telescope.

**M92** is another, slightly fainter, globular cluster. This bee-like swarm of red giants is also a great telescopic spectacle.

**FOR HOROLOGIUM, SEE PAGE 320**

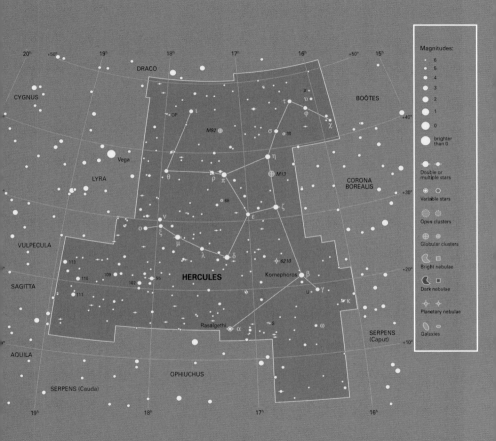

# HYDRA (THE WATER SNAKE), CORVUS, CRATER, SEXTANS

The largest constellation, although not the most exciting, Hydra consists of faint stars that straggle over a quarter of the sky. A target for Hercules, this reptile had an irksome habit: if one of its many heads was chopped off, two grew back. Hercules hacked away the heads, cauterizing the stumps with burning branches. He severed the final, immortal head and buried it, still hissing, under a stone.But the constellation is far older than this myth. Hydra is among the earliest Babylonian constellations, probably dating back to 2800 BC, when its uniquely long and straight shape marked the equator of the sky.

In the heavens, Hydra's remaining head is a pretty grouping of stars near the constellation Cancer. Its main star is **Alphard**, meaning "the solitary one."

If you have a medium-sized telescope, search out Hydra's hidden gem: the glorious face-on spiral galaxy **M83**. Nicknamed the Southern Pinwheel Galaxy, it lies near the snake's tail.

## CORVUS (THE CROW)

As far back as Babylonian times, Corvus was seen as a crow sitting on the back of the serpent, Hydra. The Greeks believed that the bird had been sent by Apollo to fetch a cup (Crater), but dallied to eat figs. It brought back the water snake (Hydra) as a scapegoat, but Apollo saw through the lie and despatched all three to the sky.

## CRATER (THE CUP)

This was the cup that a mythical crow (Corvus) was sent to fetch.

## SEXTANS (THE SEXTANT)

Polish astronomer Johannes Hevelius created this constellation in the 17th century, to commemorate the sextant he used to observe star positions.

FOR HYDRUS, SEE PAGE 360

FOR INDUS, SEE PAGE 350

FOR LACERTA, SEE PAGE 314

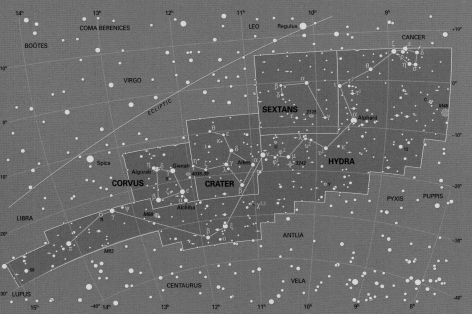

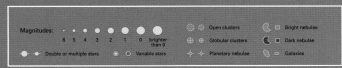

## PROMINENT STARS

| NAME | MAGNITUDE | TYPE | DISTANCE* | LUMINOSITY† | SIZE† |
|------|-----------|------|-----------|-------------|-------|
| Alphard | 2.0 | Orange giant | 177 | 780 | 50 |

*light years; †compared to the sun

# LEO (THE LION), LEO MINOR

Leo is one of the rare constellations that resemble the real thing. Lying in the Zodiac, Leo has its roots deep in prehistory. To the Greeks, it was the giant Nemean lion slaughtered by Hercules as the first of his 12 labors. The lion's flesh could not be pierced by iron, stone or bronze, so Hercules choked the beast to death.

The lion's heart is marked by brilliant **Regulus**, a star spinning so fast (completing a full rotation in just 16 hours) that its equator bulges outwards in a tangerine shape. Regulus means "little king" in Latin, and it is one of our oldest star names; over a millennium earlier, ancient Babylonians also knew this star as "the king."

The stars stretching from Regulus like a back-to-front question mark ("The Sickle") depict the lion's chest and head. Halfway along is bright **Algieba**, a lovely double star (a telescope shows an orange giant with a yellow companion). The star at the other end of Leo is appropriately called **Denebola**, "the lion's tail" in Arabic.

Just south of the main "body" of Leo are several spiral galaxies (**M65**, **M66**, **M95** and **M96**). They cannot be seen with the unaided eyes, but a sweep along the lion's tummy with a small telescope will reveal them.

Debris from Comet Tempel-Tuttle reaches the Earth each year, on or around November 17, as the **Leonid** meteor shower.

## LEO MINOR (THE LITTLE LION)

An obscure constellation introduced in the 17th century by Polish astronomer Johannes Hevelius, the little lion stands on Leo's back.

FOR LEPUS, SEE PAGE 352

## PROMINENT STARS

| NAME | MAGNITUDE | TYPE | DISTANCE* | LUMINOSITY† | SIZE† |
|------|-----------|------|-----------|-------------|-------|
| Regulus | 1.35 | Blue-white main sequence | 79 | 360 | 4 |
| Denebola | 2.1 | White main sequence | 36 | 17 | 1.7 |
| Algieba A | 2.3 | Orange giant | 131 | 285 | 29 |
| Algieba B | 3.5 | Yellow giant | 131 | 72 | 12 |

*light years, †compared to the sun

# LIBRA (THE SCALES)

From Babylonian times, astronomers have regarded this faint quadrilateral of stars near Scorpius as both the scorpion's claws *and* the scales of justice. The ancient Greeks treated Libra as part of the celestial crustacean, but Julius Caesar decided in favor of the weighing scales. When he reformed the calendar to add leap years, the Sun was in Libra when day and night were equal.

As a result, Libra is the only constellation in the Zodiac that represents an object, rather than an animal, human or god.

The Arabs gave Libra's stars delightful names. First, we have **Zubenelgenubi**, meaning "the southern claw": this white star has a fainter companion, visible with good binoculars or a small telescope. In contrast, the third-brightest star **Zubenelakrab** ("the scorpion's claw"), is an orange giant.

Brighter **Zubeneschamali** ("the northern claw") is one of the few stars that has been described as greenish in color. Astronomers normally regard stars as grading from yellow to white to blue-white, depending on their temperature, but not green. Decide for yourself, preferably using binoculars to bring out the color.

It has a fascinating history, too. Around 2,000 years ago, Greek astronomers rated Zubeneschamali as ranking with brilliant red Antares in Scorpius, but it is now clearly a lot dimmer. Perhaps it flared up in the past. The hot star is spinning 100 times faster than the Sun, and could have ejected glowing clouds of gas. It might be worth watching this star, in case it happens again.

FOR "LITTLE DIPPER," SEE URSA MINOR PAGE 364.

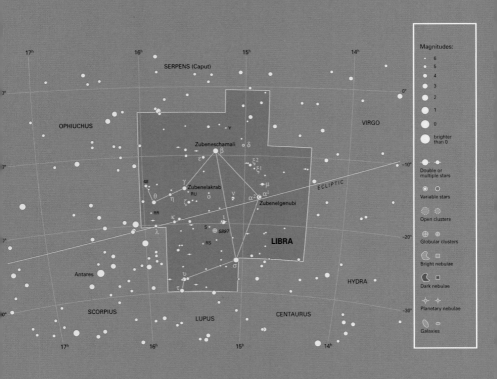

# LUPUS (THE WOLF), ARA, CIRCINUS, NORMA, TRIANGULUM AUSTRALE

For three millennia, this constellation has been seen as a fierce beast, impaled by the celestial centaur (Centaurus) and about to be sacrificed on an altar (Ara). When Renaissance scholars translated the Greek into Latin, this generic "wild thing" crystallized as a wolf. With a telescope, check out this constellation for some delightful double stars, including **Epsilon Lupi**, **Eta Lupi**, **Kappa Lupi** and **Mu Lupi**.

## ARA (THE ALTAR)

On this altar, chief Greek god Zeus made the other gods swear their loyalty, before going to war with Titans. The centaur (Centaurus) is also using the celestial altar to sacrifice the wild beast Lupus. Smoke rising from the burnt offerings appears in the sky as the Milky Way. The star cluster **NGC 6193** is the center of a large region of star formation, 4,000 light years away. A lovely sight through binoculars, and a treasure trove of stellar goodies through a telescope.

## CIRCINUS (THE COMPASSES)

Right next to the brilliant star Alpha Centauri, this unnecessary constellation was introduced by French astronomer Nicolas Louis de Lacaille in the 1750s to represent the instrument used to draw circles (as opposed to the mariner's compass, Pyxis).

## NORMA (THE CARPENTER'S SQUARE)

Another of Lacaille's constellations, created to complement the compasses (see Circinus). On sky maps, it sits uncomfortably between the sacrificial wolf (Lupus) and the altar (Ara). With binoculars or a telescope, check out the star cluster **NGC 6087**, which is centered on a pulsating Cepheid variable star.

## TRIANGULUM AUSTRALE (THE SOUTHERN TRIANGLE)

Lying not far from brilliant Alpha Centauri, this highly distinctive star triangle, with

## PROMINENT STARS

| NAME | MAGNITUDE | TYPE | DISTANCE* | LUMINOSITY† | SIZE† |
|------|-----------|------|-----------|-------------|-------|
| Atria | 1.9 | Orange giant | 390 | 5,000 | 130 |

*light years; †compared to the sun

three equal sides and three roughly equal stars, was drawn by Dutch celestial cartographer Petrus Plancius in the late 16th century. Triangulum Australe is the most easily recognisable of the "new" southern-hemisphere constellations.

FOR LYNX, SEE PAGE 290

# LYRA (THE LYRE)

A small constellation, Lyra is the early musical instrument that the god Apollo gave to Orpheus. After the death of the original musical virtuoso, the muses raised his lyre to the heavens.

Brilliant white **Vega** is the fifth-brightest star in the sky, a little fainter than Arcturus. It forms one corner of the Summer Triangle, along with Deneb (in Cygnus) and Altair (in Aquila). In 14,000 years' time, Vega will be our brilliant Pole Star. Lying only 25 light years away, Vega is surrounded by a disc of dust possibly containing baby planets.

Keen-sighted people can see the **Double-Double** (officially known as **Epsilon Lyrae**) as a pair of stars. It is easy to separate them through binoculars, and a small telescope shows each star is itself double.

**Sheliak** is a fascinating variable star. It is a very close pair of stars, almost in contact, and exchanging streamers of gas. As they orbit every 13 days, each star alternately blocks out light from the other.

The gem of Lyra lies between the two end stars, Sheliak and **Sulafat**, although you will need some telescopic power to see it clearly. The **Ring Nebula** is a wonderful example of a planetary nebula. A bit larger than Jupiter as seen in the sky, the Ring Nebula is a ghostly star corpse, a giant donut of luminous gas puffed away from a dying star about 2,000 years ago.

Each year on April 22, we're treated to a shower of shooting stars from this constellation, called the **Lyrids**. They are debris from Comet Thatcher, burning up in Earth's atmosphere.

**FOR MENSA, SEE PAGE 318**

**FOR MICROSCOPIUM, SEE PAGE 350**

**FOR MONOCEROS, SEE PAGE 298**

**FOR MUSCA, SEE PAGE 312**

**FOR NORMA, SEE PAGE 332**

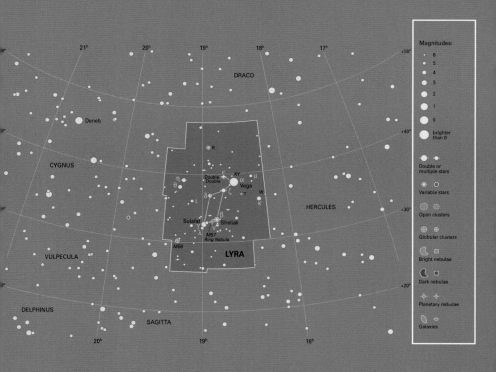

## PROMINENT STARS

| NAME | MAGNITUDE | TYPE | DISTANCE* | LUMINOSITY† | SIZE† |
|------|-----------|------|-----------|-------------|-------|
| Vega | 0.03 | White main sequence | 25 | 40 | 2.4 |

*light years; †compared to the sun

# OCTANS (THE OCTANT), APUS, CHAMAELEON, PAVO, VOLANS

The South Pole of the sky lies at Octans. In the northern hemisphere, the Earth's axis of rotation points fairly closely to the prominent star Polaris (the Pole Star). In the opposite direction, we find the very faint **Sigma Octantis**. This "southern Pole Star" is hard to spot, right on the edge of naked-eye visibility, and no use for navigation.

Nonetheless, French astronomer Nicolas Louis de Lacaille depicted these stars as a traditional navigational instrument, a reflecting octant, during his dividing up of the southern sky in the early 1750s.

### APUS (THE BIRD OF PARADISE)

The most interesting aspect of this constellation, introduced by Dutch star-map maker Petrus Plancius, is its exotic name. Scientists in Europe were amazed that these beautiful birds from the East Indies were legless ("apus" in Latin); in fact, merchants chopped off their ugly feet before sending specimens home.

### CHAMAELEON (THE CHAMELEON)

This is another constellation created by Petrus Plancius. There is little of visible importance, although dark clouds in this constellation are creating new stars.

### PAVO (THE PEACOCK)

Petrus Plancius elevated the Javan green peacock to the heavens. The brightest star, unusually, has an English name, **Peacock**, because the British Royal Air Force insisted that all navigational stars should have proper names. **NGC 6752** is a large globular cluster of stars, easily visible through binoculars.

### VOLANS (THE FLYING FISH)

The flying fish of the southern oceans were immortalized by Petrus Plancius in the 1590s. **Gamma Volantis** and **Epsilon Volantis** are lovely double stars when viewed through a small telescope.

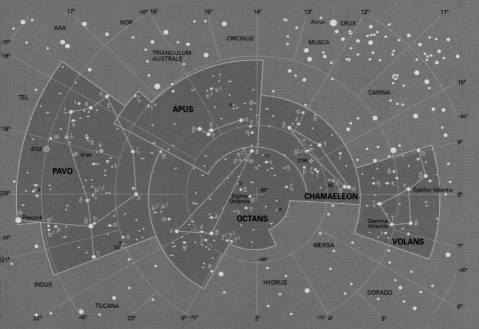

## PROMINENT STARS

| NAME | MAGNITUDE | TYPE | DISTANCE* | LUMINOSITY† | SIZE† |
|------|-----------|------|-----------|-------------|-------|
| Peacock | 1.9 | Blue-white giant | 180 | 2,200 | 5 |

*light years; †compared to the sun

# OPHIUCHUS (THE SERPENT BEARER), SERPENS

Covering a vast area of sky without any great visual drama, Ophiuchus and Serpens are two of the most ancient constellations. Representing a man entwined with a serpent, their mythology is fascinating.

To the Romans, Ophiuchus was a fabled surgeon, named Aesculapius. When he was thrown into a dungeon by King Minos of Crete for refusing to revive his dead son, Aesculapius found himself, literally, in a nest of serpents. He saw a snake arrive with a herb in its mouth, which brought a dead snake back to life. Aesculapius applied this magic balm to the dead prince, who instantly recovered. Worried that Aesculapius would make the whole human race immortal, the gods put the healer and his trusty serpent well out of the way, up in the sky.

The brightest star, **Rasalhague**, is very close to the head of another celestial giant, Hercules, and means "head of the serpent charmer." Sweep in and around Ophiuchus to spot some distant star clusters, including **M10** and **M12**, both about 15,000 light years away.

The path of the Sun, Moon and planets runs through Ophiuchus, so it is (strictly speaking) the "thirteenth sign of the Zodiac."

### SERPENS (THE SERPENT)

The snake in the tale is the only constellation that comes in two separate parts, Serpens Caput ("head") and Serpens Cauda ("tail"), separated by the body of Ophiuchus.

Each part has a treat for binoculars or, better, a telescope. The serpent's head contains **M5**, one of the largest globular star clusters in the Milky Way. In its tail, you'll find the **Eagle Nebula (M16)**. A small telescope shows the central star cluster, but it needs a bigger instrument to reveal the surrounding bright nebulosity — and the Hubble Space Telescope to lay bare the famous dark dust fingers silhouetted as the "Pillars of Creation."

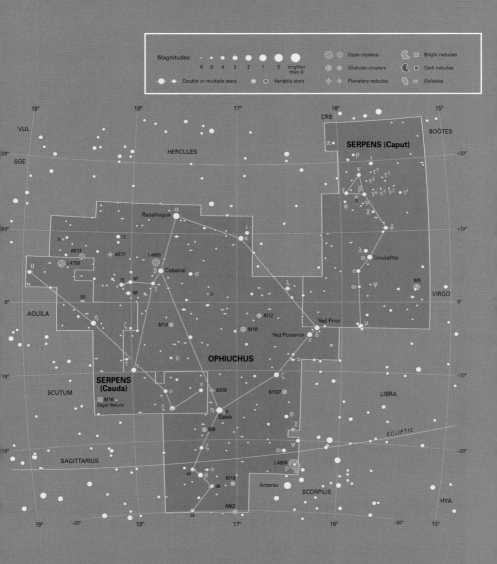

Magnitudes:

6 5 4 3 2 1 0 brighter than 0

Double or multiple stars    Variable stars

Open clusters
Globular clusters
Planetary nebulae

Bright nebulae
Dark nebulae
Galaxies

VUL

SGE

HERCULES

CRB

SERPENS (Caput)

BOÖTES

π

ρ    ι

κ    τ⁷    τ⁵    τ³

γ    β    τ⁶ τ⁵ τ⁴ τ²    τ¹

φ    ν

χ

δ

X    72

6633    6572    I.4665

I.4756

70    67

58    59    γ

η    ζ    τ

ν

μ

SERPENS (Cauda)

M16
Eagle Nebula

Rasalhague    α

ι
κ

Cebalrai    β    σ

υ

M14    M12

M10

OPHIUCHUS

ο    ν    6309

ξ    η    R    Sabik

M9

ξ

λ

Unukalhai

α

ε

ω    ψ

μ

Yed Prior

Yed Posterior    δ

υ

ζ    M107

φ

χ

ω    ψ

I.4604    ρ

Antares    SCORPIUS

M5

VIRGO

LIBRA

ECLIPTIC

AQUILA

SCUTUM

SAGITTARIUS

ο

44    θ    36    M19    45    M62

HYA

# ORION

The brilliant hunter striding across the northern winter sky is the most striking of all constellations. Fittingly, to the ancient Greeks, Orion was the most handsome man who ever lived. His deadly aim threatened to cause a total extinction of animal life, so the god Apollo sent a heavily armored scorpion that stung the hunter to death.

Diana, the goddess of the hunt, then set Orion in the sky, opposite the scorpion (the constellation Scorpius) so that as one rises the other sets. To keep Orion entertained, Diana also elevated a bull for him to fight (Taurus) and two hunting dogs (Canis Major and Canis Minor).

In the southern hemisphere, Orion signals the height of summer. Aboriginal people in Australia see the brightest stars as a canoe, with Betelgeuse and Rigel at either end; the three central stars depict the fishermen, with the fish (the misty Orion Nebula) caught in their net.

**Betelgeuse** is a red giant star, a thousand times larger than the Sun. You can spot the color easily with the naked eye. Watch carefully over several years and you may notice Betelgeuse gradually fade and brighten again, as this unstable star wobbles. It is at the end of its life, and will soon explode as a supernova, although "soon," astronomically speaking, means a few million years.

**Rigel** ("foot") forms a beautiful color contrast to Betelgeuse. Also a giant star, it is far hotter and shines with blue-white glare. A moderate telescope reveals it has a nearby companion star.

**Orion's Belt** consists of three bright stars in a straight line, a grouping that's unique in the sky. When Orion is rising, they can look like a plane coming over the horizon. Named **Alnitak** ("girdle"), **Alnilam** ("string of pearls") and **Mintaka** ("belt"), they are hot, blue-white stars, like Rigel.

**Sigma Orionis** is a lovely multiple star. A small telescope shows four of its five members. Nearby lies the silhouette of the **Horsehead Nebula**. It is beautiful

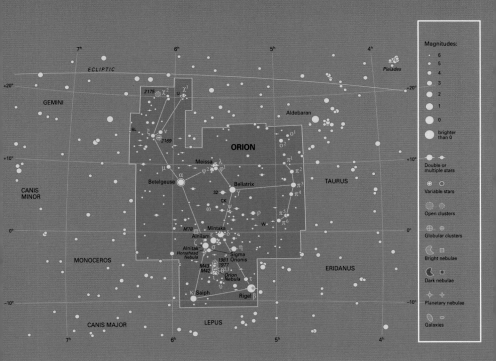

## PROMINENT STARS

| NAME | MAGNITUDE | TYPE | DISTANCE* | LUMINOSITY† | SIZE† |
|------|-----------|------|-----------|-------------|-------|
| Rigel | 0.12 | Blue-white supergiant | 860 | 125,000 | 74 |
| Betelgeuse | 0.3–1.2 | Red supergiant | 640 | 120,000 | 1,000 |
| Bellatrix | 1.6 | Blue-white giant | 250 | 6,400 | 6 |
| Alnilam | 1.7 | Blue-white supergiant | 1,300 | 375,000 | 30 |
| Alnitak | 2.0 | Blue-white supergiant | 700 | 180,000 | 19 |
| Saiph | 2.1 | Blue-white supergiant | 650 | 60,000 | 22 |
| Mintaka | 2.2 | Blue-white supergiant | 690 | 90,000 | 16 |

*light years; †compared to the sun

in photographs, but difficult to see unless you have a large telescope and dark skies.

The **Orion Nebula** is *the* glory of this already magnificent constellation. It is one of the few nebulae you can see with the naked eye, appearing as a glowing patch of gas in Orion's sword. The Orion Nebula is a lovely sight through binoculars, and a telescope reveals swirls of shining tracery around a dark gap known as the Fish's Mouth. Four stars in a tight little group here, the **Trapezium**, are the brightest of a cluster of stars that were born here only 300,000 years ago (just yesterday to astronomers).

The brightest of the Trapezium's stars, super-hot **Theta-1C Orionis**, is responsible for lighting up the surrounding gas to create the bright Orion Nebula — like a streetlight surrounding itself with a patch of illumination. The cosmic "fog" extends far beyond the Orion Nebula itself, as a dense cloud of gas and dust where massive new stars are being born right now. The thick dust hides these star embryos from ordinary optical telescopes, but they can be picked out with instruments tuned to infrared wavelengths.

On October 21–22, meteors shoot through the sky, from a point near Betelgeuse. The **Orionids** are tiny dust particles from Halley's Comet, having a moment of glory as they are destroyed in Earth's atmosphere.

FOR PAVO, SEE PAGE 336

*The magnificent Orion Nebula (lower right) is a hot-bed of star formation, a giant gas cloud heated to incandescence by energetic newborn stars. At the top of this picture, young stars are illuminating a cloud of interstellar dust as the blue-tinged nebula NGC 1977.*

# PEGASUS (THE WINGED HORSE), EQUULEUS

It may be the seventh-biggest constellation, but all that Pegasus has to offer is a large, empty square of four medium-bright stars. How *did* our ancestors manage to see an upside-down winged horse here?

In Greek myth, Pegasus sprang from the blood of Medusa the Gorgon when Perseus severed her head. However, all pre-classical civilizations had their fabled winged horses; depicted, for instance, on Etruscan and Euphratean vases.

Confusingly, **Alpheratz**, one of the stars forming **Square of Pegasus**, has been officially ceded to neighboring Andromeda.

The northernmost star of the Square, **Scheat**, is a red giant a hundred times wider than the Sun; close to the end of its life, it pulsates irregularly.

Outside the Square, we find yellow supergiant **Enif** ("nose" in Arabic). There is a faint blue star nearby, which is not associated but lies far in the background. If you tap your telescope while viewing the pair, an optical illusion makes them seem to swing like a pendulum.

Just next to Enif, lies the constellation's best-kept secret: the beautiful globular cluster **M15**. You will need a telescope for this one. M15 is 34,000 light years away and contains over 100,000 densely packed stars.

**51 Pegasi** is on the borderline of naked-eye visibility, but it achieved immortal fame in 1995 as the first star discovered to have a planet. This is classified as a "hot Jupiter": a gas giant circling 51 Pegasi far closer than Mercury orbits the Sun.

## EQUULEUS (THE LITTLE HORSE)

Although it is diminutive (only Crux is smaller) Equuleus is a venerable constellation, dating back to the ancient

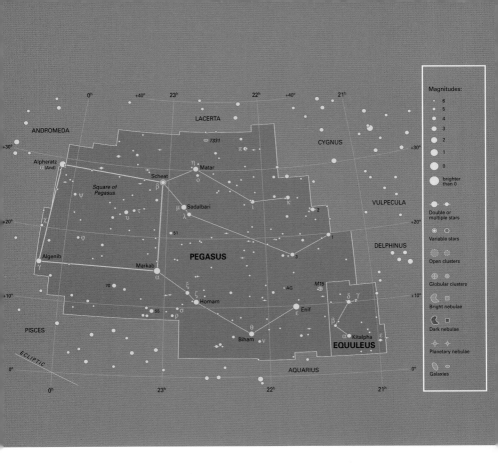

Greeks who saw it as the brother (or
son) of neighboring Pegasus. Even with
a telescope, you will not find much to
detain you here.

# PERSEUS

Perseus was a Greek hero who slew the monstrous Gorgon, Medusa, accidentally killed his grandfather with a discus and and founded the nation of the Persians. Along the way, he rescued the fair damsel Andromeda from a rampaging sea monster, for which he was awarded a place in the sky, along with Andromeda, her parents (Cassiopeia and Cepheus) and the beast (Cetus).

The brightest object in Perseus is **Mirfak** ("elbow" in Arabic), a star thousands of times more luminous than the Sun, and the senior member of the surrounding Alpha Persei star cluster (check it out with binoculars).

But the real "star" of Perseus has to be **Algol**. Representing the head of Medusa, this star regularly winks at us. In 1783, the 18-year-old John Goodricke, a profoundly deaf British amateur astronomer, correctly surmised that a fainter star was eclipsing the brighter one, once every 2 days and 21 hours, causing the brightness of Algol to drop by 70 percent.

Another gem in Perseus (or to be exact, two of them) is the **Double Cluster**. Officially named **NGC 869** and **NGC 884**, the duo lies near Cassiopeia, and is a sensational sight in binoculars. Some 7,500 light years distant, the clusters are made of bright, young, blue stars, at a mere 12 million years old.

Every August, Perseus hosts one of the most spectacular annual shooting-star displays. Radiating from a spot near the boundary with Cassiopeia, the **Perseid meteors** are debris from Comet Swift-Tuttle, burning up in the Earth's atmosphere.

FOR PHOENIX, SEE PAGE 360

FOR PICTOR, SEE PAGE 318

## PROMINENT STARS

| NAME | MAGNITUDE | TYPE | DISTANCE* | LUMINOSITY† | SIZE† |
|------|-----------|------|-----------|-------------|-------|
| Mirfak | 1.8 | Yellow-white supergiant | 560 | 7,000 | 70 |

*light years; †compared to the sun

# PISCES (THE FISHES)

Lying in a region of large, faint constellations, Pisces is no exception itself. The name is familiar only because Pisces is located in the Zodiac, the band of constellations traversed by the Sun, Moon and planets. This star pattern straggles around the sky between Aries and Aquarius, with its most distinctive feature being a ring of stars near Pegasus called **The Circlet**.

The ancient Babylonians saw two star patterns here; the stars rising towards Andromeda were the "Lady of the Heavens," while the stars near Pegasus were called the "Great Swallow." Later they became a pair of fish, denizens of a watery stream flowing from neighboring Aquarius.

In Greek mythology, the constellation was the goddess Aphrodite and her son Eros, converted into fish. They were tied together by a cord on their scaly tails, in order to escape from the monster Typhon.

In today's skies, the cord is marked by the star **Al Rischa** (meaning "cord" in Arabic). In 1779, the British astronomer William Herschel discovered that it is a double star. The two stars circle each other every 720 years; they are currently very close together, and you will need a good telescope to separate them. While you have a telescope in hand, seek out the faint spiral galaxy **M74**.

Pisces' main claim to fame is that it is the location of the Vernal Equinox, the point in the sky where the Sun crosses the celestial equator on its way from the southern to the northern hemisphere. That location, still known as the First Point of Aries, used to lie in Pisces' constellation, but, as a result of Earth's wobbling on its axis (precession), the point has now shifted into Pisces.

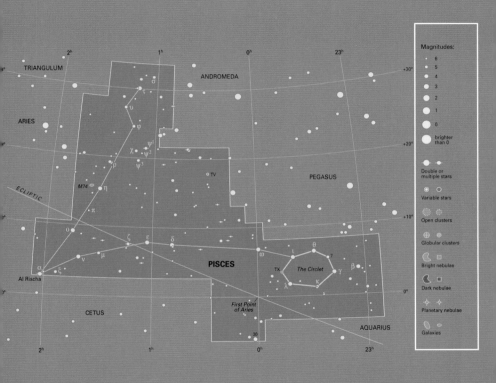

# PISCIS AUSTRINUS (THE SOUTHERN FISH), GRUS, INDUS, MICROSCOPIUM

In Greek myth, the Southern Fish is swallowing the stream of water from Aquarius' overflowing water jar. It was also the parent of Pisces, the pair of fishes in the Zodiac.

The jewel in this generally dull constellation is **Fomalhaut**, ranking among the top 20 brightest stars. The name derives from the Arabic for "mouth of the fish." For northern-hemisphere observers, Fomalhaut has the appropriate nickname: "the lonely star of Autumn." Fomalhaut isn't really lonely, though. It is orbited by a disc of debris that contains the first planet beyond the Solar System to be directly seen by our telescopes.

## GRUS (THE CRANE)

In 1597, Dutch cartographer Petrus Plancius carved off the tail of the Southern Fish (see above), to create a crane in flight in the sky.

Grus is rather more conspicuous than the other "modern" constellations in the southern sky. Look carefully and you can see that a couple of its stars are actually double, although in both cases, we are seeing stars at different distances that happen to lie in the same direction.

## INDUS (THE INDIAN)

This is another constellation created by Petrus Plancius from stars mapped by pioneering Dutch navigators. It contains nothing of interest.

## MICROSCOPIUM (THE MICROSCOPE)

In the 1750s, French astronomer Nicolas Louis de Lacaille "joined up the dots" in this barren region of sky, to create a constellation whose name reflects how little is to be found here.

FOR "PLOUGH," SEE URSA MAJOR, PAGE 362

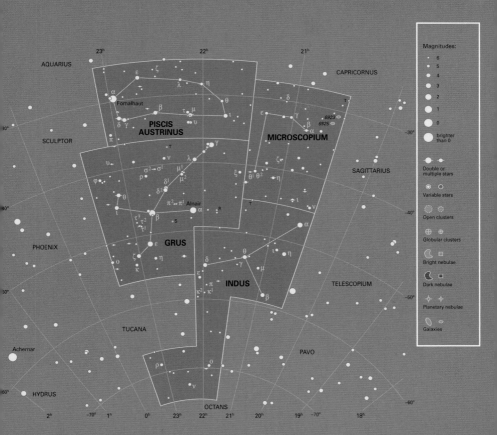

## PROMINENT STARS

| NAME | MAGNITUDE | TYPE | DISTANCE* | LUMINOSITY† | SIZE† |
|------|-----------|------|-----------|-------------|-------|
| Fomalhaut | 1.16 | White main sequence | 25 | 17 | 1.8 |
| Alnair | 1.7 | Blue-white main sequence | 101 | 380 | 3.5 |

*light years; †compared to the sun

# PUPPIS (THE STERN), CAELUM, COLUMBA, LEPUS

Part of the ancient constellation of the giant ship, *Argo* (see Vela, page 366), Puppis contains a celestial blast furnace: **Naos** is among the hottest and brightest stars known, with a temperature of 42,000°C (75,600°F), making it seven times hotter than the Sun.

Sweep the constellation with binoculars and you will stumble over several pretty star clusters, including **M46**, **M47** and **M93**. The best is **NGC 2451**, centered on a bright orange star.

## CAELUM (THE CHISEL)

French astronomer Nicolas Louis de Lacaille created this dim constellation in 1754. He depicted these stars as a pair of engraving tools, but the name is now translated as "The Chisel."

## COLUMBA (THE DOVE)

In 1592, Dutch preacher and cartographer Petrus Plancius introduced this uninteresting constellation as the dove that Noah released from the Ark to find dry land. The avian scout flies from the stern (Puppis) of the celestial ship *Argo*, which Plancius identified with Noah's Ark.

## LEPUS (THE HARE)

Crouching warily at the feet of Orion, we find the hare, safe, for the moment at least, as the giant hunter engages the ferocious bull, Taurus. Lepus was first mentioned by the Greek astronomer Ptolemy, in around AD 150, when he drew up the original list of 48 constellations.

**Gamma Leporis** is faint, but worth looking at through binoculars or a small telescope, as it is a pretty double star. You could also track down the variable **Hind's Crimson Star** (usually just too faint to be seen with the naked eye), one of the reddest stars known; 19th-century British astronomer John Russell Hind described it as "like a drop of blood on a black field."

FOR PYXIS, SEE PAGE 366

FOR RETICULUM, SEE PAGE 318

## PROMINENT STARS

| NAME | MAGNITUDE | TYPE | DISTANCE* | LUMINOSITY† | SIZE† |
|------|-----------|------|-----------|-------------|-------|
| Naos | 2.2 | Blue supergiant | 1,090 | 550,000 | 14 |

*light years; †compared to the sun

# SAGITTARIUS (THE ARCHER), CORONA AUSTRALIS, TELESCOPIUM

From Babylonian times, the star pattern of Sagittarius has represented an archer, in the form of a centaur; with the torso of a man and the body of a horse. A distinct curve of three stars marks his bent bow, while the point of his arrow is aimed at Scorpius, the fearsome celestial scorpion.

From the northern hemisphere, the shape looks more like a teapot, with the handle to the left, and the spout (to the right) issuing the steamy clouds of the Milky Way.

The center of our galaxy lies in Sagittarius, although it is totally obscured by dark clouds. Nonetheless, the Milky Way here is rich in nebulae and star clusters. The wonderful **Lagoon Nebula** is visible to the naked eye on clear nights, and is magnificent through a small telescope. You will need a telescope to see its three-lobed neighbor, the **Trifid Nebula**.

Sweep through Sagittarius with binoculars for more fantastic sights. Near the border with Aquila, you will find a bright patch of stars in the Milky Way (**M24**). Nearby, binoculars reveal another fine star-forming region, the **Omega Nebula**, although you will need a telescope to discern its distinctive arch shape. Finally, the fuzzy patch **M22** is a globular cluster: a swarm of almost a million stars, lying 10,000 light years away.

## CORONA AUSTRALIS (THE SOUTHERN CROWN)

Originally seen by the Greeks as a wreath, or as the points of Sagittarius' five arrows, this distinct circlet of stars does not have a lot to offer the amateur astronomer.

## TELESCOPIUM (THE TELESCOPE)

Created by the French astronomer Nicolas Louis de Lacaille in the 1750s, this puny constellation certainly requires a powerful telescope to discern anything of interest.

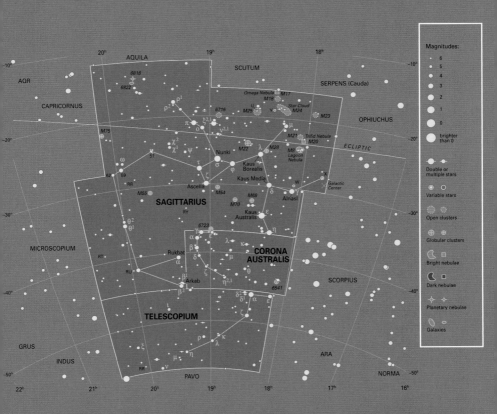

| PROMINENT STARS | | | | | |
| NAME | MAGNITUDE | TYPE | DISTANCE* | LUMINOSITY† | SIZE† |
| Kaus Australis | 1.8 | Blue-white giant | 140 | 360 | 7 |
| Nunki | 2.1 | Blue-white main sequence | 230 | 3,300 | 5 |

*light years; †compared to the sun

# SCORPIUS (THE SCORPION)

This constellation is intimately linked to Orion, on the opposite side of the sky. When a scorpion killed the mighty hunter, the gods set the two opponents as far apart as possible, so that Orion sets as Scorpius rises.

Scorpius is a rare constellation that looks like its namesake, with claws stretching out towards Libra, and a fine tail with a distinctive "sting." Although it is famous as one of the signs of the Zodiac, the path of the Sun, Moon and planets actually just clips Scorpius and mainly runs through Ophiuchus.

At the scorpion's heart lies brilliant red giant **Antares**, "the rival of Mars," whose ruddiness even surpasses the Red Planet. This huge star would stretch to the asteroid belt, if placed in our Solar System. Its fainter (although still 170 times brighter than the Sun) blue-white companion is visible in a moderate telescope and looks greenish against Antares' strong red hue.

The sting is marked by a pair of bright, but unrelated, stars, **Shaula** and **Lesath**, nicknamed the "cat's eyes." Here you will also find three fine star clusters that are visible to the naked eye: **M6**, **M7** and **NGC 6231**. A telescope reveals their stars clearly, and gives great views of the double star **Graffias**, the multiple-star **Jabbah** and the striking "double-double" **Xi Scorpii**.

Finally, use binoculars to find **M4**, close to Antares. This swarm of tens of thousands of stars is the nearest of our Galaxy's giant globular clusters; even so, it lies 7,200 light years away.

FOR SCULPTOR, SEE PAGE 310

FOR SCUTUM, SEE PAGE 286

FOR SERPENS, SEE PAGE 338

FOR SEXTANS, SEE PAGE 326

FOR "SOUTHERN CROSS," SEE CRUX, PAGE 312

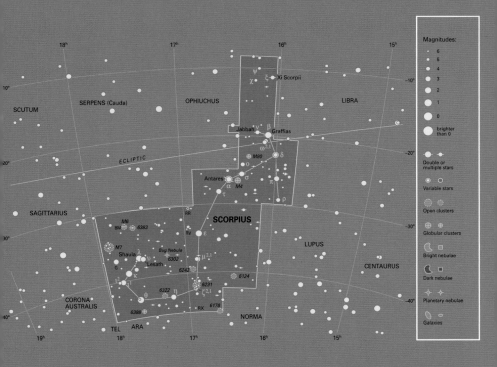

## PROMINENT STARS

| NAME | MAGNITUDE | TYPE | DISTANCE* | LUMINOSITY† | SIZE† |
|------|-----------|------|-----------|-------------|-------|
| Antares | 0.96 | Red supergiant | 550 | 57,500 | 880 |
| Shaula | 1.6 | Blue-white giant | 500 | 30,000 | 7 |
| Lesath | 2.7 | Blue-white giant | 580 | 12,300 | 6 |

*light years; †compared to the sun

# TAURUS (THE BULL)

One of the oldest constellations, Taurus dates back to the ancient Babylonians. They called it the "Bull in Front," because the Sun passed through on the Vernal Equinox.

Taurus is dominated by **Aldebaran**, the bull's baleful blood-red eye. Like other aging red giant stars, Aldebaran varies slightly in brightness. The surrounding "head" is formed by the **Hyades** star cluster. This is just a chance alignment, though; the Hyades lie over twice as far away, at 153 light years.

Vying for attention are the Seven Sisters: the **Pleiades** star cluster. Despite the name, skywatchers see any number from six to eleven stars, depending on their eyesight. The Pleiades is glorious through binoculars or a small telescope.

Taurus has two stars as "horns": **El Nath** (Arabic for "the butting one") to the north and **Zeta Tauri** (with the tongue-twisting Babylonian name Shurnarkabtishashutu, "star in the bull towards the south"). Near the latter lies the **Crab Nebula**, the wreck of a star that has blown itself apart. Chinese astronomers witnessed this supernova in AD 1054, and today the debris is visible through a medium-sized telescope. Radio astronomers have found a pulsar in its center: the rapidly spinning core of the dead star.

FOR TELESCOPIUM, SEE PAGE 354

FOR TRIANGULUM, SEE PAGE 288

FOR TRIANGULUM AUSTRALE, SEE PAGE 332

FOR TUCANA, SEE PAGE 360

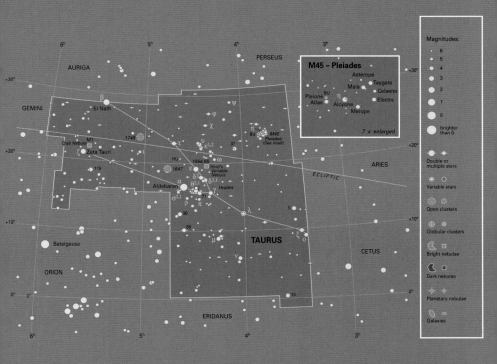

## PROMINENT STARS

| NAME | MAGNITUDE | TYPE | DISTANCE* | LUMINOSITY† | SIZE‡ |
|------|-----------|------|-----------|-------------|-------|
| Aldebaran | 0.85 | Orange giant | 65 | 520 | 44 |
| El Nath | 1.7 | Blue-white giant | 130 | 700 | 4 |

*light years;  †compared to the sun

# TUCANA (THE TOUCAN), HYDRUS, PHOENIX

After Dutch navigators Pieter Keyser and Frederick de Houtman measured the positions of stars in the previously uncharted southern skies, in 1597, their compatriot Petrus Plancius arranged them into 12 new constellations, including Tucana.

To the naked eye, **Beta Tucanae** is a double star, while a small telescope shows the brighter star itself is actually double. Powerful instruments reveal all three are close doubles, making Beta Tucanae a six-star system.

Turn binoculars on the faint "star" **47 Tucanae**, and you will see that it is a luminous ball with a bright center. The second-brightest globular cluster, 47 Tucana is a giant swarm of a million stars 17,000 light years away.

But the glory of Tucana is easily visible to the naked eye: the glowing mass of the **Small Magellanic Cloud**. The petite sister of the Large Magellanic Cloud in Dorado, it is one of our nearest neighbors, 200,000 light years away. Sweep it with binoculars or a telescope for a grandstand view of star clusters and nebulae.

### HYDRUS (THE LITTLE WATER SNAKE)

Just to confuse everyone, the sky has two water snakes. Vast Hydra is one of the oldest constellations, representing a mythical beast. Hydrus, on the other hand, was dreamt up by Petrus Plancius to depict the real ocean-dwelling snakes of the South Seas.

### PHOENIX (THE PHOENIX)

The largest of the new southern constellations introduced by Petrus Plancius, this is the mythical bird that rises from its own ashes. **Zeta Phoenicis** is an unusual quadruple star. A small telescope shows one companion, while a more powerful telescope reveals another star, closer in. But the main "star" actually consists of a very close star pair, alternately

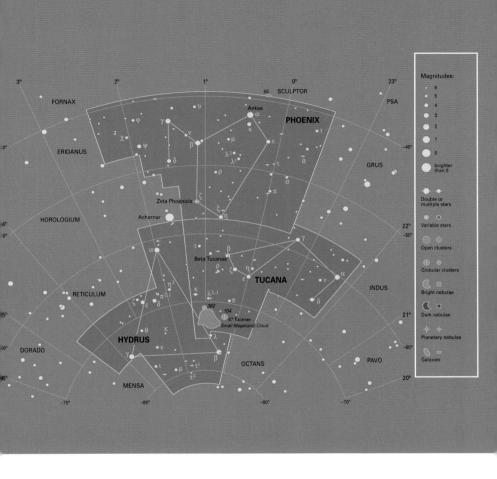

eclipsing each other every 1.7 days: it's
a southern counterpart to the famous
eclipsing binary Algol, in Perseus.

# URSA MAJOR (THE GREAT BEAR), CANES VENATICI

Possibly the oldest constellation, Ursa Major dates back a staggering 30,000 years. The native North Americans, who left the Old World before the last Ice Age, see the same bear-shape in this star pattern as the ancient peoples of Siberia and Europe.

In Greek myth, Zeus seduced the nymph Callisto, who bore a son, Arcas. Not amused, Hera turned Callisto into a bear. Years later, Arcas encountered the bear while hunting. To prevent matricide, Zeus flung Callisto into the sky by her stumpy tail, which inevitably stretched, to become the Great Bear.

The seven brightest stars form a familiar pattern, known variously as **The Plough** and the **Big Dipper**. The two Pointers — **Merak** and **Dubhe** — signpost the way to the Pole Star.

Unusually, Ursa Major contains a double star you can split with the naked eye; **Mizar**, in the middle of the tail, has a fainter companion, **Alcor**. A telescope reveals Mizar itself is a close double.

Still with a telescope, check out the striking galaxy pair **M81** and **M82**. The **Pinwheel Galaxy (M101)** is stunning in long-exposure images, but through the telescope looks merely large and dim.

## CANES VENATICI (THE HUNTING DOGS)

Created by the 17th-century Polish astronomer Johannes Hevelius, these two hunting dogs eternally chase the Great Bear. The brighter star is known as **Cor Caroli**, Charles's Heart, after the executed British king, Charles I.

A telescope reveals three of the top 20 brightest galaxies in this obscure constellation: the spirals **M94** and **NGC 4258 (M106)**, along with the **Whirlpool Galaxy (M51)**, a spiral galaxy whipped up by a smaller passing galaxy.

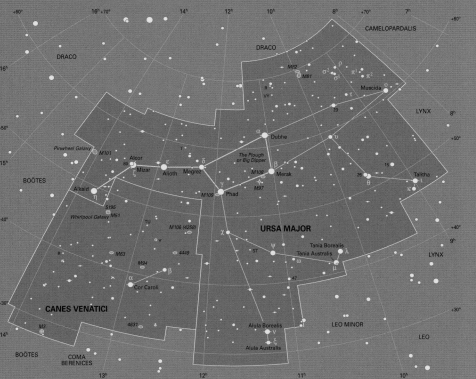

## PROMINENT STARS

| NAME | MAGNITUDE | TYPE | DISTANCE* | LUMINOSITY† | SIZE† |
|------|-----------|------|-----------|-------------|-------|
| Alioth | 1.8 | White main sequence | 82 | 108 | 4 |
| Dubhe | 1.8 | Orange giant | 123 | 300 | 30 |
| Alkaid | 1.8 | Blue-white main sequence | 104 | 1,300 | 3 |

*light years; †compared to the sun

# URSA MINOR (THE LITTLE BEAR), DRACO

Ursa Minor is a miniature version of Ursa Major, its brightest stars being known as the **Little Dipper** in North America. To the ancient Greeks, the Little Bear was, fittingly, the son of the Great Bear. As the human boy Arcas was about to slay his mother, a nymph who had been turned into a bear, Zeus changed Arcas into the same form, too, and flung them both into the sky for safety.

**Polaris** lies almost exactly above Earth's north pole. We spin "underneath" it, so the Pole Star is almost a fixed point in the sky, always due north, making it very handy for navigation. But the Earth's axis wobbles in space every 26,000 years (an effect known as precession) so Polaris has only been our Pole Star since the Middle Ages, and will eventually lose that title again.

Through a small telescope, you can see that Polaris is double. The main star is a Cepheid variable, changing slightly in brightness as it pulsates in and out.

### DRACO (THE DRAGON)

Writhing between the two bears in the northern sky, the cosmic dragon is associated with Hercules' 12 labors: the monster's head rests on the (upside-down) superhero's feet.

**Thuban** makes up for its faintness with fame. Because the Earth's axis wobbles gradually (see Polaris, left), Thuban was our Pole Star when the Great Pyramids were built in Egypt. Around 2800 BC, Thuban actually lay closer to the celestial pole than Polaris does now.

Another star with a hidden claim to fame is **Eltanin**. Currently 148 light years away, in 1.5 million years' time Eltanin will zoom past us at just 28 light years' distance, to become the brightest star in the sky.

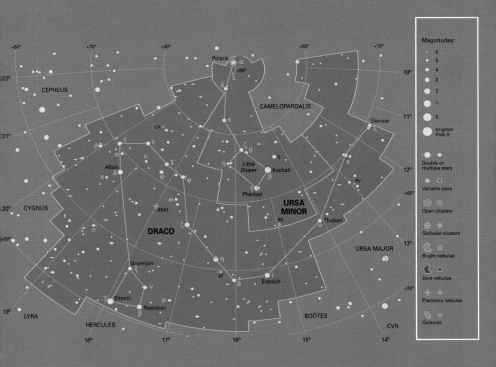

## PROMINENT STARS

| NAME | MAGNITUDE | TYPE | DISTANCE* | LUMINOSITY† | SIZE† |
|------|-----------|------|-----------|-------------|-------|
| Polaris | 2.0 | Yellow-white supergiant | 325 | 2,500 | 46 |
| Kochab | 2.1 | Orange giant | 130 | 390 | 42 |

*light years; †compared to the sun

# VELA (THE SAILS), ANTLIA, PYXIS

The great ship *Argo* that carried Jason and the Argonauts on their legendary quest for the Golden Fleece was placed in the sky as the huge constellation Argo Navis. The 18th-century French astronomer Nicolas Louis de Lacaille split this unwieldy star pattern into its keel (Carina), stern (Puppis) and the billowing sails (Vela).

Lying largely in the Milky Way, Vela is rich in bright stars, nebulae and star clusters. Two of its stars, **Markeb** and **Delta Velorum**, make up the **False Cross** with Avior and Aspidiske in Carina (marked on the chart of Carina, page 303).

**Suhail** (**Gamma Velorum**) is an amazing star. It is a delightful double in binoculars, and a small telescope reveals two extra fainter companions. The brightest component actually comprises two massive stars too close to split with a telescope. The brighter star is a supergiant; the other is the brightest example of a Wolf-Rayet star, a searingly hot giant that has lost its outer layer of gases. Both will explode as supernovae.

With a good telescope you can see that **Delta Velorum** is a fine double star. The brighter component is an eclipsing binary with a period of 45 days and slightly brighter than Algol.

## ANTLIA (THE AIR PUMP)

Introduced by Lacaille, to commemorate the air pump devised by French inventor Denis Papin, the constellation is largely a vacuum for amateur astronomers. (If Lacaille had immortalized another of Papin's inventions, we could have had the constellation of the Pressure Cooker!)

## PYXIS (THE COMPASS)

Lacaille created this constellation to represent a mariner's compass near the stern of the ship *Argo*. It doesn't point the way to anything interesting.

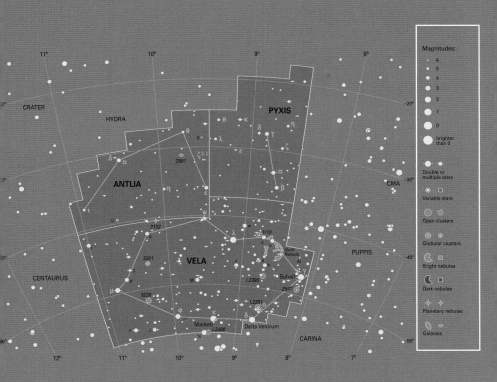

## PROMINENT STARS

| NAME | MAGNITUDE | TYPE | DISTANCE* | LUMINOSITY† | SIZE† |
|------|-----------|------|-----------|-------------|-------|
| Suhail | 1.8 | Blue supergiant | 840 | 150,000 | 13 |
| Delta Velorum | 2.0 | White main sequence | 81 | 56 | 2.6 |

*light; years;  †compared to the sun

# VIRGO (THE VIRGIN), COMA BERENICES

Y-shaped **Virgo** is the second largest constellation, and among the oldest: it originally depicted the Earth Mother. During the northern harvest time the Sun passes through Virgo, but it needs a bit of imagination to make out a virtuous maiden holding an ear of corn (Spica).

**Spica** is a brilliant hot star, boasting a temperature of 40,350°F (22,400°C). It has a close companion, and the two stars inflict a heavy gravitational toll on each other, distorting each other into egg-shapes.

Named after a goddess of childbirth, **Porrima** is a delightful double star as seen through a medium telescope. To the south, a telescope reveals the striking shape of the **Sombrero Galaxy**.

The glory of Virgo lies in the "bowl." Scan it with a small telescope, and you will fall over myriad faint, fuzzy blobs, just a few of the 2,000 galaxies making up the gigantic **Virgo Cluster**. The brightest of them is **M49**, while the central giant elliptical galaxy, **M87**, is a powerful source of radio waves and harbors a truly massive black hole.

## COMA BERENICES (BERENICE'S HAIR)

Unlike most constellations, the main group of stars, which looks like a large fuzzy patch, forms a real cluster in space. Greek astronomer Ptolemy saw it as the tuft on the tail of Leo, the lion. Its present name is more romantic: Egyptian queen Berenices promised to sacrifice her beautiful hair if her husband returned successfully from battle in 243 BC. He did. She carried out her vow and the gods placed her flowing locks in the sky.

A telescope reveals the **Blackeye Galaxy (M64)**, a lovely sight, despite its name, and some outlying galaxies of the neighboring **Virgo Cluster** (see above).

FOR VOLANS, SEE PAGE 336

FOR VULPECULA, SEE PAGE 316

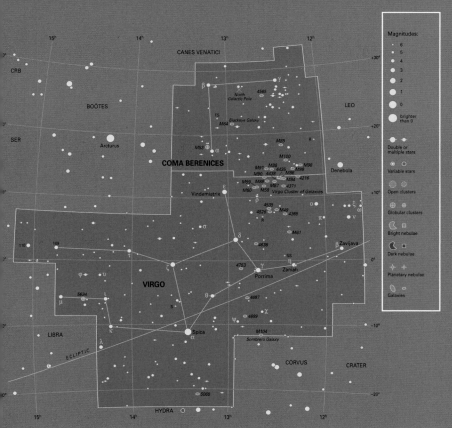

| 15ʰ | | 14ʰ | | 13ʰ | | 12ʰ |

CANES VENATICI

CRB

BOÓTES

SER

Arcturus

COMA BERENICES

Vindemiatrix

VIRGO

LIBRA

ECLIPTIC

Spica

CORVUS

CRATER

HYDRA

North Galactic Pole

4565

Blackeye Galaxy

M64

M53

M85

M100

M91 M88
M90 4435 M99
M98

4438 M86
M59 M88 M84 4216
M60 M58 M87 4371
Virgo Cluster of Galaxies

4535
4526 M49
R 4365

M61

4636

SS

Zaniah

4753

Porrima

4697

4699

M104
Sombrero Galaxy

5068

LEO

Denebola

Zavijava

5634

CRITICAL MARKERS: γ β α δ ε ζ η θ ι κ λ μ ν σ τ φ ψ χ ω ο π

**Magnitudes:**

- 6
- 5
- 4
- 3
- 2
- 1
- 0
- brighter than 0

Double or multiple stars

Variable stars

Open clusters

Globular clusters

Bright nebulae

Dark nebulae

Planetary nebulae

Galaxies

+30°
+20°
+10°
0°
-10°
-20°

## PROMINENT STARS

| NAME | MAGNITUDE | TYPE | DISTANCE* | LUMINOSITY† | SIZE† |
|------|-----------|------|-----------|-------------|-------|
| Spica | 1.04 | Blue-white main sequence | 260 | 12,100 | 7 |

*light years; †compared to the sun

# CHAPTER 10
# REFERENCE

# TOP 30 SKY SIGHTS

| NAME | TYPE | CONSTELLATION | BEST SEEN* | PAGE |
|------|------|---------------|------------|------|
| Large Magellanic Cloud | Irregular/spiral galaxy | Dorado | January | 258, 318 |
| Orion Nebula, M42 | Nebula | Orion | January | 216, 340 |
| Sirius | Brightest star | Canis Major | February | 212, 296 |
| M41 | Star cluster | Canis Major | February | 296 |
| Praesepe, M44 | Star cluster | Cancer | March | 222, 294 |
| Southern Pleiades | Star cluster | Carina | April | 220, 302 |
| Carina Nebula | Nebula | Carina | April | 218, 302 |
| Southern Cross | Asterism** | Crux | May | 312 |
| Coalsack | Dark cloud | Crux | May | 214, 312 |
| Jewel Box | Star cluster | Crux | May | 312 |
| The Plough | Asterism** | Ursa Major | May | 362 |
| Mizar/Alcor | Double star | Ursa Major | May | 224, 362 |
| Omega Centauri | Globular cluster | Centaurus | May | 306 |
| Alpha Centauri | Double/nearest star | Centaurus | June | 306 |
| M13 | Globular cluster | Hercules | July | 324 |
| Xi Scorpii | Double-double star | Scorpius | July | 356 |
| NGC 6231 | Star cluster | Scorpius | July | 356 |
| M7 | Star cluster | Scorpius | August | 356 |
| Lagoon Nebula, M8 | Nebula | Sagittarius | August | 219, 354 |
| Omega Nebula, M17 | Nebula | Sagittarius | August | 354 |
| Epsilon Lyrae | Double-double star | Lyra | August | 334 |
| Albireo | Double star | Cygnus | August | 224, 314 |
| Andromeda Galaxy, M31 | Spiral galaxy | Andromeda | October | 260, 282 |
| 47 Tucanae | Globular cluster | Tucana | November | 360 |

| NAME | TYPE | CONSTELLATION | BEST SEEN* | PAGE |
| --- | --- | --- | --- | --- |
| Small Magellanic Cloud | Irregular galaxy | Tucana | November | 258, 360 |
| Double Cluster | Star clusters | Perseus | November | 346 |
| Algol | Variable star | Perseus | December | 226, 346 |
| Pleiades, M45 | Star cluster | Taurus | December | 220, 358 |
| Milky Way | Galaxy | All around the sky | February, August | 254–7 |
| Polaris | Pole Star | Ursa Minor | All year | 212, 364 |

*Month when the object is highest in the sky at 10 p.m.; it will usually be visible for a few months around this time.
**A striking star-pattern

*The Large Magellanic Cloud*

# BRIGHTEST STARS

## THE FIRST-MAGNITUDE STARS

| | NAME | MAGNITUDE | CONSTELLATION | PAGE |
|---|---|---|---|---|
| 1 | Sirius | -1.47 | Canis Major | 212, 240, 296 |
| 2 | Canopus | -0.72 | Carina | 212, 302 |
| 3 | Alpha Centauri | -0.27 | Centaurus | 306 |
| 4 | Arcturus | -0.04 | Boötes | 234, 292 |
| 5 | Vega | 0.03 | Lyra | 212, 334 |
| 6 | Capella | 0.08 | Auriga | 290 |
| 7 | Rigel | 0.12 | Orion | 212, 340 |
| 8 | Procyon | 0.34 | Canis Minor | 298 |
| 9 | Betelgeuse | 0.42 | Orion | 212, 234, 340 |
| 10 | Achernar | 0.50 | Eridanus | 320 |
| 11 | Hadar | 0.60 | Centaurus | 306 |
| 12 | Altair | 0.77 | Aquila | 286 |
| 13 | Acrux | 0.77 | Crux | 312 |
| 14 | Aldebaran | 0.85 | Taurus | 358 |
| 15 | Antares | 0.96 | Scorpius | 212, 235, 356 |
| 16 | Spica | 1.04 | Virgo | 212, 228, 368 |
| 17 | Pollux | 1.15 | Gemini | 322 |
| 18 | Fomalhaut | 1.16 | Piscis Austrinus | 231, 350 |
| 19 | Deneb | 1.25 | Cygnus | 212, 314 |
| 20 | Mimosa | 1.25 | Crux | 312 |
| 21 | Regulus | 1.35 | Leo | 328 |

*The Southern Cross (Crux) with the dark Coalsack, and Hadar on the left: this picture contains three of the 21 first-magnitude stars.*

# NEAREST STARS

| NAME | DISTANCE* | TYPE | LUMINOSITY† | NOTES |
|---|---|---|---|---|
| Proxima Centauri | 4.24 | Red main sequence | 0.0017 | Discovered in 1915 |
| Alpha Centauri A | 4.37 | Yellow main sequence | 1.5 | Third-brightest star |
| Alpha Centauri B | 4.37 | Orange main sequence | 0.5 | May have planet |
| Barnard's Star | 5.96 | Red main sequence | 0.0035 | Named for E. E. Barnard |
| Luhman 16 | 6.59 | Brown dwarf‡ pair | 0.00001 | Discovered in 2013; named for Kevin Luhman |
| WISE 0855-0714 | 7.18 | Brown dwarf‡ | 0.00001 | Discovered in 2014 |
| Wolf 359 | 7.78 | Red main sequence | 0.0011 | Named for Max Wolf |
| Lalande 21185 | 8.29 | Red main sequence | 0.025 | Named for Jerome Lalande |
| Sirius A | 8.58 | White main sequence | 25.4 | Brightest star |
| Sirius B | 8.58 | White dwarf | 0.026 | "The Pup" |
| UV Ceti | 8.73 | Red main sequence | 0.00005 | Double star |
| Ross 154 | 9.68 | Red main sequence | 0.0038 | Powerful flares |
| Ross 248 | 10.32 | Red main sequence | 0.0018 | Voyager 2 will pass close by |
| Epsilon Eridani | 10.52 | Orange main sequence | 0.34 | Has nearest confirmed extrasolar planet |
| Lacaille 9352 | 10.74 | Red main sequence | 0.033 | Named for Nicholas Louis de Lacaille |
| Ross 128 | 10.92 | Red main sequence | 0.0035 | Named for Frank Ross |
| WISE 1506+7027 | 11.09 | Brown dwarf‡ | 0.00001 | Named for Wide-field Infrared Survey Explorer |
| EZ Aquarii | 11.27 | Red main sequence | 0.00008 | Triple star |
| Procyon A | 11.40 | Yellow-white main sequence | 6.9 | Eighth-brightest star |
| Procyon B | 11.40 | White dwarf | 0.0005 | Predicted by its gravity |

| NAME | DISTANCE* | TYPE | LUMINOSITY† | NOTES |
|------|-----------|------|-------------|-------|
| 61 Cygni | 11.41 | Orange main sequence | 0.15 | Double star; first star to have distance measured |
| Struve 2398 | 11.53 | Red main sequence | 0.039 | Double star; named for Friedrich Struve |
| Grcombridge 34 | 11.62 | Red main sequence | 0.0064 | Double star; named for Stephen Groombridge |
| Epsilon Indi | 11.82 | Orange main sequence | 0.22 | Nearest star similar to Sun |
| DX Cancri | 11.83 | Red main sequence | 0.00065 | Flare star |
| Tau Ceti | 11.89 | Yellow main sequence | 0.52 | May have five planets |

*light years; †compared to the Sun; ‡failed star

*A small telescope reveals nearby Alpha Centauri as a beautiful double star.*

# TOP 20 BRIGHTEST GALAXIES

| NAME | MAGNITUDE | DISTANCE* | TYPE |
|---|---|---|---|
| Large Magellanic Cloud | 0.4 | 0.16 | Irregular galaxy |
| Small Magellanic Cloud | 2.2 | 0.20 | Irregular galaxy |
| Andromeda Galaxy, M31 | 3.4 | 2.5 | Giant spiral galaxy |
| Triangulum Galaxy, M33 | 5.7 | 2.9 | Spiral galaxy |
| Centaurus A | 6.8 | 12 | Giant elliptical galaxy |
| M81 | 6.9 | 12 | Spiral galaxy |
| Silver Coin Galaxy, NGC 253 | 7.1 | 11 | Spiral galaxy, with starburst center |
| Southern Pinwheel, M83 | 7.5 | 16 | Face-on spiral galaxy |
| Pinwheel Galaxy, M101 | 7.9 | 24 | Giant spiral galaxy |
| NGC 55 | 7.9 | 6.7 | Edge-on spiral |
| Sombrero Galaxy, M104 | 8.0 | 30 | Edge-on spiral galaxy with prominent equatorial dust lane |
| M94 | 8.2 | 15 | Face-on spiral galaxy |
| NGC 4258 | 8.4 | 24 | Spiral galaxy |
| M82 | 8.4 | 12 | Edge-on starburst spiral galaxy |
| M49 | 8.4 | 50 | Giant elliptical galaxy |
| Whirlpool Galaxy, M51 | 8.4 | 26 | Spiral galaxy with companion |
| NGC 2403 | 8.5 | 11 | Spiral galaxy |
| NGC 1291 | 8.5 | 33 | Spiral galaxy with outer ring |
| Blackeye Galaxy, M64 | 8.5 | 16 | Spiral galaxy with thick dust lane |
| M87 | 8.6 | 53 | Giant elliptical galaxy |

| MASS[†] | CONSTELLATION | PAGE |
|---|---|---|
| 10 billion | Dorado | 258, 318 |
| 6.5 billion | Tucana | 258, 360 |
| 1,200 billion | Andromeda | 260, 282 |
| 50 billion | Triangulum | 288 |
| 1,000 billion | Centaurus | 264, 306 |
| 200 billion | Ursa Major | 362 |
| 100 billion | Sculptor | 262, 310 |
| 200 billion | Hydra | 263, 326 |
| 1,000 billion | Ursa Major | 362 |
| 20 billion | Sculptor | 310 |
| 1,000 billion | Virgo | 368 |
| 60 billion | Canes Venatici | 362 |
| 190 billion | Canes Venatici | 362 |
| 50 billion | Ursa Major | 362 |
| 200 billion | Virgo | 368 |
| 160 billion | Canes Venatici | 266, 362 |
| 100 billion | Camelopardalis | 304 |
| 150 billion | Eridanus | 320 |
| 40 billion | Coma Berenices | 368 |
| 6,000 billion | Virgo | 264, 368 |

*millions of light years; †Suns

# CONSTELLATION LIST

| CONSTELLATION | MEANING | GENITIVE* | ABBR. | AREA DEG.²† | PAGE |
|---|---|---|---|---|---|
| Andromeda | Andromeda (Princess) | Andromedae | And | 722 | 282 |
| Antlia | Air Pump | Antliae | Ant | 239 | 366 |
| Apus | Bird of Paradise | Apodis | Aps | 206 | 336 |
| Aquarius | Water Carrier | Aquarii | Aqr | 980 | 284 |
| Aquila | Eagle | Aquilae | Aql | 652 | 286 |
| Ara | Altar | Arae | Ara | 237 | 332 |
| Aries | Ram | Arietis | Ari | 441 | 288 |
| Auriga | Charioteer | Aurigae | Aur | 657 | 290 |
| Boötes | Herdsman | Boötis | Boo | 907 | 292 |
| Caelum | Chisel | Caeli | Cae | 125 | 352 |
| Camelopardalis | Giraffe | Camelopardalis | Cam | 757 | 304 |
| Cancer | Crab | Cancri | Cnc | 506 | 294 |
| Canes Venatici | Hunting Dogs | Canum Venaticorum | CVn | 465 | 362 |
| Canis Major | Great Dog | Canis Majoris | CMa | 380 | 296 |
| Canis Minor | Little Dog | Canis Minoris | CMi | 183 | 298 |
| Capricornus | Sea Goat | Capricorni | Cap | 414 | 300 |
| Carina | Keel | Carinae | Car | 494 | 302 |
| Cassiopeia | Cassiopeia (Queen) | Cassiopeiae | Cas | 598 | 304 |
| Centaurus | Centaur | Centauri | Cen | 1,060 | 306 |
| Cepheus | Cepheus (King) | Cephei | Cep | 588 | 308 |
| Cetus | Sea Monster | Ceti | Cet | 1,231 | 310 |
| Chamaeleon | Chameleon | Chamaeleontis | Cha | 132 | 336 |
| Circinus | Compasses | Circini | Cir | 93 | 332 |
| Columba | Dove | Columbae | Col | 270 | 352 |

| CONSTELLATION | MEANING | GENITIVE* | ABBR. | AREA DEG.²† | PAGE |
|---|---|---|---|---|---|
| Coma Berenices | Berenice's Hair | Comae Berenices | Com | 386 | 368 |
| Corona Australis | Southern Crown | Coronae Australis | CrA | 128 | 354 |
| Corona Borealis | Northern Crown | Coronae Borealis | CrB | 179 | 292 |
| Corvus | Crow | Corvi | Crv | 184 | 326 |
| Crater | Cup | Crateris | Crt | 282 | 326 |
| Crux | Southern Cross | Crucis | Cru | 68 | 312 |
| Cygnus | Swan | Cygni | Cyg | 804 | 314 |
| Delphinus | Dolphin | Delphini | Del | 189 | 316 |
| Dorado | Goldfish | Doradus | Dor | 179 | 318 |
| Draco | Dragon | Draconis | Dra | 1,083 | 364 |
| Equuleus | Little Horse | Equulei | Equ | 72 | 344 |
| Eridanus | River | Eridani | Eri | 1,138 | 320 |
| Fornax | Furnace | Fornacis | For | 398 | 320 |
| Gemini | Twins | Geminorum | Gem | 514 | 322 |
| Grus | Crane | Gruis | Gru | 366 | 350 |
| Hercules | Hercules (Hero) | Herculis | Her | 1,225 | 324 |
| Horologium | Clock | Horologii | Hor | 249 | 320 |
| Hydra | Water Snake | Hydrae | Hya | 1,303 | 326 |
| Hydrus | Little Water Snake | Hydri | Hyi | 243 | 360 |
| Indus | Indian | Indi | Ind | 294 | 350 |
| Lacerta | Lizard | Lacertae | Lac | 201 | 314 |
| Leo | Lion | Leonis | Leo | 947 | 328 |
| Leo Minor | Little Lion | Leonis Minoris | LMi | 232 | 328 |
| Lepus | Hare | Leporis | Lep | 290 | 352 |

| CONSTELLATION | MEANING | GENITIVE* | ABBR. | AREA DEG.²† | PAGE |
|---|---|---|---|---|---|
| Libra | Scales | Librae | Lib | 538 | 330 |
| Lupus | Wolf | Lupi | Lup | 334 | 332 |
| Lynx | Lynx | Lyncis | Lyn | 545 | 290 |
| Lyra | Lyre | Lyrae | Lyr | 286 | 334 |
| Mensa | Table Mountain | Mensae | Men | 153 | 318 |
| Microscopium | Microscope | Microscopii | Mic | 210 | 350 |
| Monoceros | Unicorn | Monocerotis | Mon | 482 | 298 |
| Musca | Fly | Muscae | Mus | 138 | 312 |
| Norma | Carpenter's Square | Normae | Nor | 165 | 332 |
| Octans | Octant | Octantis | Oct | 291 | 336 |
| Ophiuchus | Serpent Bearer | Ophiuchi | Oph | 948 | 338 |
| Orion | Orion (Hunter) | Orionis | Ori | 594 | 340 |
| Pavo | Peacock | Pavonis | Pav | 378 | 336 |
| Pegasus | Pegasus (Winged Horse) | Pegasi | Peg | 1,121 | 344 |
| Perseus | Perseus (Hero) | Persei | Per | 615 | 346 |
| Phoenix | Phoenix | Phoenicis | Phe | 469 | 360 |
| Pictor | Painter's Easel | Pictoris | Pic | 247 | 318 |
| Pisces | Fishes | Piscium | Psc | 889 | 348 |
| Piscis Austrinus | Southern Fish | Piscis Austrini | PsA | 245 | 350 |
| Puppis | Stern | Puppis | Pup | 673 | 352 |
| Pyxis | Compass | Pyxidis | Pyx | 221 | 366 |
| Reticulum | Reticle | Reticuli | Ret | 114 | 318 |
| Sagitta | Arrow | Sagittae | Sge | 80 | 316 |
| Sagittarius | Archer | Sagittarii | Sgr | 867 | 354 |

| CONSTELLATION | MEANING | GENITIVE* | ABBR. | AREA DEG.²† | PAGE |
|---|---|---|---|---|---|
| Scorpius | Scorpion | Scorpii | Sco | 497 | 356 |
| Sculptor | Sculptor | Sculptoris | Scl | 475 | 310 |
| Scutum | Shield | Scuti | Sct | 109 | 286 |
| Serpens | Serpent | Serpentis | Ser | 637 | 338 |
| Sextans | Sextant | Sextantis | Sex | 314 | 326 |
| Taurus | Bull | Tauri | Tau | 797 | 358 |
| Telescopium | Telescope | Telescopii | Tel | 252 | 354 |
| Triangulum | Triangle | Trianguli | Tri | 132 | 288 |
| Triangulum Australe | Southern Triangle | Trianguli Australis | TrA | 110 | 332 |
| Tucana | Toucan | Tucanae | Tuc | 295 | 360 |
| Ursa Major | Great Bear | Ursae Majoris | UMa | 1,280 | 362 |
| Ursa Minor | Little Bear | Ursae Minoris | UMi | 256 | 364 |
| Vela | Sails | Velorum | Vel | 500 | 366 |
| Virgo | Virgin | Virginis | Vir | 1,294 | 368 |
| Volans | Flying Fish | Volantis | Vol | 141 | 336 |
| Vulpecula | Little Fox | Vulpeculae | Vul | 268 | 316 |

*The Latin "possessive" form of the constellation name, used in conjunction with Greek letters to name stars. For instance, the brightest star in Centaurus is called Alpha Centauri, meaning "alpha *of* the Centaur."; †Square degrees; for comparison, the area of the whole sky is 41,253 square degrees.

# PLANET-SPOTTING 2015–20

Mercury and Venus are visible in the evening or dawn sky, during the periods listed in the tables below. Mars, Jupiter and Saturn move right round the sky: the tables show the constellations in which they are located, by month. Check when these constellations, and their planetary inhabitants, are visible to you using the star charts on pages 28–35.

## MERCURY

|  | MORNING | EVENINWG |
|---|---|---|
| 2015 | February–March, June, October | January, May, August–September, December |
| 2016 | January–February | April |
| May 9, 2016: Mercury transits the Sun | — | — |
| 2016 | May–June, September | August, December |
| 2017 | January, May, September, December | March–April, July–August, November |
| 2018 | January, April–May, August December, | March, June–July, November |
| 2019 | March–April, August | February, June, October |
| November 11, 2019: Mercury transits the Sun | — | — |
| 2019 | November–December |  |
| 2020 | March–April, July, November | February, May–June, September–October |

# Venus

| | MORNING | EVENING |
|---|---|---|
| 2015 | August–December | January–August |
| 2016 | January–April | July–December |
| 2017 | March–November | January–March |
| 2018 | October–December | February–October |
| 2019 | January–June | September–December |
| 2020 | June–December | January–May |

# Mars

| | | CONSTELLATION* |
|---|---|---|
| 2015 | January | Aquarius |
| | February–March | Pisces |
| | August | Cancer |
| | September–October | Leo |
| | November–December | Virgo |
| 2016 | January–February | Libra |
| | March–May | Scorpius/Ophiuchus |
| | June–July | Libra |
| | August–September | Scorpius/Ophiuchus |
| | October | Sagittarius |
| | November–December | Capricornus |
| 2017 | January | Aquarius |
| | February | Pisces |
| | March | Aries |
| | April–May | Taurus |
| | September | Leo |
| | October–December | Virgo |
| 2018 | January | Libra |
| | February | Scorpius/Ophiuchus |
| | March–May | Sagittarius |
| | June–October | Capricornus |
| | November–December | Aquarius |

## Mars (con't)

| | | CONSTELLATION* |
|---|---|---|
| 2019 | January | Pisces |
| | February–March | Aries |
| | April–May | Taurus |
| | June | Gemini |
| | October–November | Virgo |
| | December | Libra |
| 2020 | January | Scorpius/Ophiuchus |
| | February–March | Sagittarius |
| | April | Capricornus |
| | May–June | Aquarius |
| | July–December | Pisces |

## Jupiter

| | | CONSTELLATION* |
|---|---|---|
| 2015 | January | Leo |
| | February–May | Cancer |
| | June–July | Leo |
| | October–December | Leo |
| 2016 | January–August | Leo |
| | November–December | Virgo |
| 2017 | January–September | Virgo |
| | November–December | Libra |
| 2018 | January–October | Libra |
| | December | Scorpius/Ophiuchus |
| 2019 | January–November | Scorpius/Ophiuchus |
| 2020 | January–December | Sagittarius |

## Saturn

| | | CONSTELLATION* |
|---|---|---|
| 2015 | January | Libra |
| | February–May | Scorpius/Ophiuchus |
| | June–October | Libra |
| | December | Scorpius/Ophiuchus |
| 2016 | January–November | Scorpius/Ophiuchus |

## Saturn (con't)

| | | CONSTELLATION* |
|---|---|---|
| 2017 | January–February | Scorpius/Ophiuchus |
| | March–May | Sagittarius |
| | June–November | Scorpius/Ophiuchus |
| 2018 | January–December | Sagittarius |
| 2019 | February–December | Sagittarius |
| 2020 | February–March | Sagittarius |
| | April–June | Capricornus |
| | July–December | Sagittarius |

## Uranus

| | | CONSTELLATION* |
|---|---|---|
| 2015 | January–December | Pisces |
| 2016 | January–December | Pisces |
| 2017 | January–December | Pisces |
| 2018 | January–April | Pisces |
| | May–December | Aries |
| 2019 | January–December | Aries |
| 2020 | January–December | Aries |

## Neptune

| | | CONSTELLATION* |
|---|---|---|
| 2015 | January–December | Aquarius |
| 2016 | January–December | Aquarius |
| 2017 | January–December | Aquarius |
| 2018 | January–December | Aquarius |
| 2019v | January–December | Aquarius |
| 2020 | January–December | Aquarius |

*The planets move along the constellations of the Zodiac, but the true astronomical star-patterns differ from the pseudo-scientific astrological "signs of the Zodiac." In particular, the feet of Ophiuchus (the Serpent Bearer) take up most of the region between Libra and Sagittarius, with Scorpius (the Scorpion) contributing just one small claw.

# SOLAR ECLIPSES 2015–20

## 2015

### MARCH 20

- Total eclipse along narrow track through north Atlantic Ocean, including Faeroe Islands and Svalbard
- Partial eclipse in Europe, North Africa and north-west Asia

### SEPTEMBER 13

- Partial eclipse in South Africa, south Indian Ocean and part of Antarctica

## 2016

### MARCH 9

- Total eclipse along narrow track through Indonesia and north Pacific Ocean
- Partial eclipse in Australia, south-east Asia and north Pacific Ocean

### SEPTEMBER 1

- Annular eclipse along narrow track through equatorial Africa and Madagascar
- Partial eclipse in most of Africa and Indian Ocean

## 2017

### FEBRUARY 26

- Annular eclipse along narrow track through southern Chile and Argentina, south Atlantic Ocean and south-west Africa
- Partial eclipse in southern South America, all of south Atlantic Ocean and most of western Africa

### AUGUST 21

- Total eclipse along narrow track through USA
- Partial eclipse in all of North America, central America, Caribbean, north Atlantic Ocean and northern South America

## 2018

### FEBRUARY 15

- Partial eclipse in southern South America and Antarctica

### JULY 13

- Partial eclipse between Australia and Antarctica

## AUGUST 11

- Partial eclipse in Scandinavia, Greenland, Arctic Ocean and Russia

# 2019

## JANUARY 6

- Partial eclipse in China, Japan and north Pacific Ocean

## JULY 2

- Total eclipse along narrow track through south Pacific Ocean, Chile and Argentina
- Partial eclipse in most of south Pacific Ocean and South America

## DECEMBER 26

- Annular eclipse along narrow track through Saudi Arabia, Sri Lanka and Indonesia
- Partial eclipse in India, Indian Ocean, China, south-east Asia and northern Australia

# 2020

## JUNE 21

- Annular eclipse along narrow track through north-east Africa, Saudi Arabia, north India and China
- Partial eclipse in all of north-east Africa, east Europe, north Indian Ocean and most of Asia

## DECEMBER 14

- Total eclipse along narrow track through south Pacific Ocean, Chile, Argentina and south Atlantic Ocean
- Partial eclipse in east Pacific Ocean, southern South America, all of South Atlantic Ocean and part of Antarctica

# LUNAR ECLIPSES

## 2015

### APRIL 4

- Total eclipse, visible from west North America, Pacific Ocean, Australasia and eastern Asia

### SEPTEMBER 28

- Total eclipse, visible from Europe, Africa, Atlantic Ocean, the Americas and east Pacific Ocean

## 2016

—

## 2017

### AUGUST 7

- Partial eclipse, visible from Europe, Africa, east Atlantic Ocean, Asia, Indian Ocean, Australasia and west Pacific Ocean

## 2018

### JANUARY 31

- Total eclipse, visible from most of Asia, Indian Ocean, Australasia, Pacific Ocean and most of North America

### JULY 27

- Total eclipse, visible from Europe, Africa, south Atlantic Ocean, Indian Ocean, most of Asia, western Australasia

## 2019

### JANUARY 21

- Total eclipse, visible from the Americas, east Pacific Ocean, Atlantic Ocean, Europe, west Africa

### JULY 16

- Partial eclipse, visible from South America, Atlantic Ocean, Europe, Africa, most of Asia, Indian Ocean and Australasia

## 2020

—

# RESOURCES/ FIND OUT MORE

## PLANETARIUM APPS

Find out what in the sky right now, with an app for your phone or tablet. Point your device towards the sky and it will identify stars and planets for you.

*Google Sky Map* for Android phones and tablets

*Star Chart* for iPhone and iPad

## PLANETARIUM SOFTWARE

To find out what the sky looks like overhead tonight — or for any other time and place on Earth — on your computer, you can use planetarium programs. You can buy high-powered planetarium software, but the following are among the freeware available:

*Computer Aided Astronomy*
http://www.astrosurf.com/c2a/english/

*Skychart/ Cartes du Ciel*
http://www.ap-i.net/skychart/en/start

*Stellarium*
http://www.stellarium.org/

*WorldWide Telescope*
http://www.worldwidetelescope.org/

## TELESCOPE MANUFACTURERS

*APM Telescopes*
Refractors, reflectors
http://www.apm-telescopes.de/en/
Telescopes-.html

*Celestron*
Binoculars, catadioptrics, CCD cameras
http://www.celestron.com/

*Meade Instruments*
Catadioptrics, apochomatic refractors, solar telescopes
http://www.meade.com/

*Sky-Watcher Telescopes*
Specializes in Dobsonians; also refractors, reflectors
http://www.skywatcher.com/index.php

*Takahashi*
Fluorite apochromatic refractors
Japanese homepage: http://www.takahashijapan.com/index.html

Takahashi America: http://takahashiamerica.com/

Takahashi Europe: http://www.takahashi-europe.com/en/index.php

## IMAGING DEVICES

*Atik Cameras*
CCD cameras for imaging faint nebulae and galaxies
http://www.atik-cameras.com/

*Point Grey Research*
The ultra-lightweight Flea3 webcam-type imager gives sharp views of the planets
http://ww2.ptgrey.com/USB3/Flea3

*Starlight Xpress*
CCD cameras for imaging faint nebulae and galaxies
http://www.sxccd.com/

## REMOTE OBSERVING

*iTelescope*
Over a dozen telescopes around the world, including Australia, Spain and New Mexico
http://www.itelescope.net/

*MicroObservatory*
Operated by NASA and the Harvard-Smithsonian Center for Astrophysics, with free access to five telescopes across the USA
http://mo-www.cfa.harvard.edu/OWN/index.html

*MyAstroPic*
Telescopes in the UK, USA, Spain and Chile
http://astro.wdg.com.ua/

*Slooh*
Telescope in Canary Islands; lots of community projects
http://events.slooh.com/

## DARK-SKY SITES

*International Dark-Sky Association*
The arbiter for dark-sky sites and international organization for combating light pollution
http://darksky.org/

## RADIO ASTRONOMY

*Society of Amateur Radio Astronomers*
Advice and forum for backyard radio
astronomy
http://radio-astronomy.org/

## CITIZEN SCIENCE

*BOINC (Berkeley Open Infrastructure for
Network Computing)*
On a site originally devised to support
SETI@home (see below), you can now
let your computer automatically help to
map the Milky Way galaxy, search for
gravitational waves and determine the
structure of the Universe (as well as many
other science projects)
http://boinc.berkeley.edu/

*SETI@home*
Let your computer automatically trawl
data for the first call from ET
http://setiathome.berkeley.edu/

*SETILive*
Actively search for radio signals from
extraterrestrial life, using data from the
pioneering Allan Telescope Array
http://www.setilive.org/

*Zooniverse*
Take an active role in exploring galaxies,
Mars, the Moon, solar explosions, black
holes and planets orbiting other stars, all
from the comfort of your own chair
https://www.zooniverse.org/

## MOON

View the current phase of the Moon
https://www.fourmilab.ch/cgi-bin/Earth/
action?opt=-m&img=MoonTopo.evif

## PLANETS

Download NASA's free World Wind
software to explore interactive globes of
the Earth, Moon, Mars and Jupiter, as well
as Jupiter's satellites.
http://worldwind.arc.nasa.gov/java/

## SOLAR ACTIVITY/AURORA ALERT

Check out sunspots and aurora
predictions
http://spaceweather.com/

## ECLIPSES

*Eclipses of the Moon*
Follow up our list of dates with detailed
predictions from NASA
eclipse.gsfc.nasa.gov/lunar.html

*Eclipses of the Sun*
Follow up our list of dates with detailed
predictions from NASA
eclipse.gsfc.nasa.gov/solar.html

## SATELLITES

Find out when the International Space
Station and other bright satellites are
passing over your viewing location:
http://www.heavens-above.com/

## ORGANIZATIONS TO JOIN

*Society for Popular Astronomy*
For beginners of all ages
www.popastro.com

*British Astronomical Association*
For experienced observers in the UK and
abroad
www.britastro.org

*American Association for Variable Star
Observers*
Specialises in observations of variable
stars
www.aavso.org

*Royal Astronomical Society of Canada*
For astronomers in Canada
www.rasc.ca

*Royal Astronomical Society of New
Zealand*
For astronomers in New Zealand
www.rasnz.org.nz

*Astronomical Society of Australia*
Website has links to astronomy societies
throughout Australia
astronomy.org.au/amateur/

# INDEX

## A

achromatic refractors 40, 42–3
active galaxies 252, 268
Albireo 224–5, 314, 315
Alcor 224, 225
Aldebaran 13, 222, 358, 359
Aldrin, Buzz 66, 84, 85
Algol 226, 227
Allen, Paul 232–3
Almaaz 226–7
Almach 282
Alpha Centauri 208, 209, 306, 307,377
Alphard 326, 327
Alpheratz 280, 344
Andromeda 282-3
    Galaxy 20, 260–1, 264, 270, 282
Antares 31, 212, 235, 356
Antikythera mechanism 12, 13
Antiope 151
Antlia 366
Apus 336
Aquarius 284–5
Aquila 286–7
Ara 332
Arcturus 31, 32, 234
Ariel 137
Aries 207, 288
asteroid belt 90, 147, 152
asteroids 146–52, 154–5
    Late Heavy Bombardment 78–9, 92
    and the Moon 76, 78–9
    moons 148
    naming 147
astronomy holidays 57
astrophotography 49, 52–5
Auriga 290–1
aurorae 8, 27, 200–1

## B

Beehive Cluster 294
Bell, Jocelyn 246–7
Beta Pictoris 230
Betelgeuse 206–7, 212, 234, 340, 341
Big Bang 9, 20, 252, 272–3, 274
binoculars 24, 36–7
black holes 9, 19, 21, 248–9, 252
    and quasars 269
Bok Globules 219
Boötes 292
Brahe, Tycho 14, 15, 159, 242, 304
brown dwarfs 228

## C

Caelum 352
calendar 184–5
Callisto 121, 122
cameras 52–5, 392
Cancer (the Crab) 294
    Praesepe 222–3, 294
Canes Venatici 362
Canis Major 296–7, 340
Canis Minor 298–9, 340
Canopus 32
Capella 290
Capricornus 300–1
carbonaceous chondrites 155
Carina 220, 302–3
    Nebula 218–19
Cassiopeia 32, 242, 279, 304–5
catadioptric telescopes 48–51
CCDs (charge-coupled devices) 55
celestial latitude/longitude 280–1
Centaurus 264–5, 306–7
Cepheid variables 236–7, 259, 364
Cepheus 308
Ceres 146, 147, 148, 150, 156

Cetus 237, 310–11
Chamaelon 336
Charon 157
Chelyabinsk asteroid 152, 153, 155
Chinese constellations 279
Circlet 280, 348
citizen science 62–3, 393
Clark, Alvin 240
Clerke, Agnes 166
Coalsack 214, 312, 375
Columba 352
Coma Berenices 267, 368
comets 8, 90–1, 145, 158–65
    and meteors 166
    nuclei 162–3, 165
    observing 164
    tails 160, 161
    see also Halley's Comet
constellations 10, 11, 13, 24, 26,
        276–369
    Chinese 279
    list of 380–3
    mapping the sky 280–1
    seasonal charts 28–35
Copernicus, Nicolaus 14, 15, 20, 184
Corona Australis 354
Corona Borealis 292
Corvus 267, 326
Crab Nebula 246–7, 358
Crater 326
Cygnus 28, 206, 214, 224–5, 314–15
    A 268, 269
    X-1 248–9

## D

dark adaptation 27
dark energy 20, 275
dark matter 62, 273

dark-sky sites 57, 204, 392
Darquier, Antoine 238
deep-sky objects 55
Delphinus 316
Delta Cephei 236
Deneb 206, 212, 314, 315
Doppler Shift 272–3
Dorado 318
Draco 364
Drake, Frank 232
Dumbbell Nebula 238–9

### E

Earth 12, 92, 102–3
   and Mars 105, 112
   and the Moon 66–7, 70, 70–1, 86
   orbiting the Sun 14, 15, 16
   reflectivity (albedo) 67
   tides 70
   and Venus 98, 99
eclipsing binaries 226–7
Ecliptic 281
elliptical galaxies 252, 261, 264–5, 268
Enceladus 67, 130–1
Epsilon Aurigae 226–7
equinoxes 187
Equuleus 344–5
Eridanus 253, 320, 321
Eris 156
Eta Carinae 219, 302
Europa 115, 120, 121, 122, 233
Evans, Robert 243

### F

Fabricius, David 237
fireballs 166
Firework Nebula 241
Fornax 320

### G

galaxies 20–1, 252–71

brightest 378–9
clusters 270
interacting 266–7
   Local Group 270, 288
photographing 55
viewing 50
Galileo Galilei 16–17, 20, 40, 78
   and the Milky Way 254–5
   and the planets 120, 123, 126
   and Praesepe 222
gamma rays 60
Ganymede 120, 121, 122
Gaspra 148, 149
Gegenschein 172
Gemini 222, 322–3
Geminid meteors 322
Greek astronomers 12–13, 14, 206
Gregorian calendar 184–5
Grus 350

### H

Hale-Bopp Comet 159, 161, 164, 165
Halley's Comet 17, 158, 159,
   nucleus 162
   and shooting stars 169
Hay, Will 132, 134
Henderson, Thomas 208
Hercules 324–5
Herschel, William 17, 20, 131, 136, 262
Hipparchus 204
Horologium 320
Horsehead Nebula 214, 215, 218, 342
Hubble, Edwin 20, 272–3
Hubble Space Telescope 9, 38,157,
   204, 275
   Cepheids 236
   Orion Nebula 217
   Proxima Centauri 229
Huygens, Christiaan 126
Hyades cluster 222, 358
Hydra 326–7
Hydrus 360

Hyperion 131

### I

Iapetus 131, 134
Ida 148, 150
Indus 350
infrared telescopes 60
International Space Station 103,
   176, 394
Io 53, 120–1, 122
Iridium flares 177
ISON Comet 160, 162

### J

James Webb Space Telescope 38
Jupiter 24, 45, 90, 92–3, 114–25
   and asteroids 146, 147
   Great Red Spot 114, 116–17,
     119, 124, 125
   map 116–19
   missions to 115
   moons 53, 93, 115, 120–1, 124
   observing 122–5, 386
   photographing 53
   radio waves from 59
   and Saturn 126

### K

Keck telescopes 45, 47
Kepler-186 230–1
Khayyam, Omar 172, 184, 185
Kuiper belt 90, 139, 156

### L

Lacerta 314
Lagoon Nebula 219
Large Magellanic Cloud (LMC)
   244, 245, 259–60,318, 373
leap years 184
Leo 28, 31, 32, 222, 279, 328–9
Lepus 352

Libra 330, 330–1
light pollution 57, 204
Lovell Radio Telescope 58
Lupus 332
Lynx 290
Lyra 28, 334–5

## M

M87 galaxy 264, 270
Magellanic Clouds 20
Main Sequence stars 228–9
Mars 15, 90, 92, 104–13
   and asteroids 146, 147
   life on 106–7, 233
   moons 92, 104, 113
   observing 112–13, 385–6
Mayor, Michael 230
Mensa 318
Mercury 90, 92, 94–7, 384
meteorites 152, 154–5
meteors (shooting stars) 8, 166–71
   showers 168–9, 171
the Mice 267
Microscopium 350
Milky Way 9, 17, 20, 91, 172, 252,
   254–9
   and the Andromeda Galaxy 260–1
   black hole 269
   and comets 160
   companion galaxies 258–9
   and Cygnus 214
   galaxy cluster 270
   globular clusters 256
   photographing 55
   radio waves 59
Mimas 131
Mira 237, 310
Miranda 136–7
Mizar 224, 225, 362
Monoceros 298
Moon 8, 11, 16, 64–87
   astronauts 84

birth of 84
   craters 67, 78–81, 83
   and the Earth 66–7, 70, 86
   eclipses 12, 86–7, 390, 394
   far side 71
   Full Moon 71, 72, 73
   lava and volcanoes 82–3
   Man in the Moon 66, 72, 74, 75
   map 72–3
   mountains and 'seas' 72–3, 74–5,
      76–7
   orbit 69, 70–1
   phases 12, 24, 68–9
   photographing 52, 53, 55
   reflectivity 67
   and solar eclipses 198–9
   viewing 72, 74–5
morning skies 29–30
Muralis Quadrans 169
Musca 312

## N

near-Earth objects (NEOs) 152–3
nebulae 50, 55, 216–19, 259
   Crab Nebula 246–7, 358
   photographing 55
   planetary 238–9
   viewing 50
Neptune 24, 39, 90, 92–3, 138–9
   moons 93, 138–9
   observing 140–1, 387
neutron stars 19, 247
Newton, Isaac 16–17, 44, 159
Newtonian reflector 46
NGC galaxies 264, 267, 304, 310,
   320
noctilucent clouds 174–5
Norma 332
northern hemisphere 28–32, 187
Northern Lights 8, 27, 201
Nova Persei 241

## O

Octans 336
Omega Centauri 256, 306
Oort cloud 90, 160
Ophiuchus 338
Orion 27, 28, 31, 32, 96, 205, 212,
   279, 340–3
   Molecular Cloud 216–18
   Nebula 205, 210, 216–17, 257,
      259, 342, 343

## P

Pavo (Peacock) 336, 337
Pegasus 344
Penzias, Arno 20
Perseid meteor shower 169
Perseus 226, 227, 346–7
Persian astronomers 13, 14,
   206, 226
Phoenix 360–1
Pisces 280, 348–9
   Austrinus 350–1
planetarium software 150, 391
planetary nebulae 238–9
planets 8, 12, 88–141, 210
   birth of 144, 145
   dwarf planets 148, 150, 156
   extrasolar 230–3
   gas giants 92–3, 222
   life on other planets 232–3
   moons 92, 93
   photographing 53, 55
   planet-spotting 384–7
   rocky 92
planispheres 26, 27
Pleiades 13, 220, 221, 222, 358
Plough 224, 225, 280, 362
Pluto 90, 156–7
Polaris 212, 364, 365
Praesepe 222–3, 294
Procyon 298, 299

protostars 214
Proxima Centauri 18, 212, 229
Psyche 149, 150
Ptolemy, Claudius 12, 13, 14
pulsars 9, 59, 246–7
Puppis 352
Pyxis 366

## Q

quasars 9, 21, 59, 252, 268–9
Queloz, Didier 230

## R

radio astronomy 58–9, 232–3, 269, 393
radio galaxies 252, 268, 269
radio waves 58–9, 60
red giants 13, 210, 234–5
reflecting telescopes 44–7
refractors 40–3
remote observing 56–7, 392
Riccioli, Giovanni Battista 74, 78
Rigel 212, 340
Ring Nebula 238

## S

Sagitta 316
Sagittarius 255, 257, 354–5
satellites 61, 176–7, 195, 394
Saturn 43, 92–3, 126–35
    moons 67, 93, 128–9, 130–1, 134
    observing 132–5, 386–7
    rings 126, 128–9, 134
Schmidt telescopes 48–50, 51
Scorpius 28, 32, 212, 235, 279, 356–7
Sculptor 310
Sculptor galaxy 262–3
Scutum 286
seasonal charts 24
seasons 186–7
Serpes 338
SETI Institute 232–3

Seven Sisters see Pleiades
Sextans 326
Shakespeare, William 69, 158
Shoemaker-Levy 9 (comet) 115
shooting stars see meteors
    (shooting stars)
Sirius (Dog Star) 204, 206, 212, 296,
    297
    the Pup 240, 298
61 Cygni 208, 209
Small Magellanic Cloud (SMC)
    258–9, 360
Solar Dynamics Observatory 195
Solar System 12, 17, 62, 90–3
    birth 144
    see also planets
South Pole 32
Southern Cross (Crux) 28, 32, 214,
    279, 312–13, 375
southern hemisphere 32–5, 187
Southern Lights 8, 27, 201
Southern Pinwheel 263
Southern Pleiades 220–2, 302
spectroscopy 19, 213
Spica 210, 212, 228–9, 368, 369
spiral galaxies 252, 253, 260–3
stars 18–19, 90, 202–49
    birth of 60, 62–3, 210, 214–15
    brightest 374–5
    clusters 220–3
    colours 212–13
    distances 208–10
    double stars 224–5, 226
    eclipsing binaries 226–7
    extrasolar planets 230–3
    life cycle 210
    magnitude system 204
    Main Sequence 228–9
    naming 206–7
    nearest 376–7
    variable 236–7
Stonehenge 11, 187

summer solstice 11, 187
Sun 8, 11, 16, 178–99, 210, 211,
    224, 228, 241
    and the 24-hour day 183
    atmosphere 194–5
    and aurorae 200–1
    birth 144
    and the calendar 184–5
    centre 180
    eclipses 188, 198–9, 388–9, 394
    internal structure 190–1
    and Mercury 94, 96
    and the Moon 68–9, 76
    observing 188–9
    orbit 184
    and planetary nebulae 238
    and the planets 14, 15, 90, 91, 93
    radio waves 58, 59, 61, 62
    and the seasons 186–7
    size 180
    solar storms 196, 201
    sunrise and sunset 187, 197
    sunspots 193, 196–7
    and supernovae 244
    surface 192–3
    and Venus 98–9, 100–1
sundials 183
supernovae (exploding stars)
    9, 19, 59, 210, 242–5
    and black holes 248–9
    and neutron stars 247
    Southern Pinwheel 263

## T

Tarantula Nebula 259, 318
Taurus 222, 246, 358–9
telescopes 17, 18, 24–5, 38–51
    and astrophotography 49, 53–5
    and the atmosphere 38
    buying 39, 391–2
    compact 38–9
    eyepieces and magnification 43

finders 45
GO TO 49, 141
light grasp 39
magnification 39
mountings 49
radio astronomy 58–9, 232–3
reflectors 44–7
refractors 40–3
remote observing 56
size 39
solar 189
Telescopium 354
tides 70
Titan 130, 131, 134
Titania 137
Trapezium cluster 216
Triangulum 288
Australe 332–3
galaxy 261–2, 270
Triton 139
Tucana 360–1

## U
ultraviolet rays 60, 62

Unbriel 137
Universe
expansion 252, 274
structure 20–1
Uranus 17, 24, 39, 90, 92–3, 136–7
moons 93, 136–7, 141
and Neptune 138
observing 140–1, 387
Ursa Major 32, 169, 206, 279, 280, 362–3
Ursa Minor 364

## V
variable stars 236–7
Vega 208, 209, 212, 334, 335
Vela 366–7
Venus 24, 90, 92, 98–101
observing 100–1, 385
Vesta 148, 149, 150
Virgo 210, 368–9
Cluster 270, 368
VLT Survey Telescope 39
Volcans 336
Vulpecula 238–9, 316–17

## W
webcams 55
Welles, Orson 106
Wells, H.G. 106
Whirlpool galaxy 266, 362
white dwarfs 9, 210, 240–1
exploding 242, 244
Wild Duck Cluster 286
Wilson, Robert 20
winter solstice 11, 187

## X
X-rays 60, 61, 248, 249

## Y
Yerkes Telescope 41, 43

## Z
Zodiac 24, 281
Zodiac Light 172–3

# ACKNOWLEDGMENTS

**Editorial Director:** Trevor Davies

**Editor:** Pollyanna Poulter

**Copy Editor:** Cathy Lowne

**Proofreaders:** Jane Birch and Mandy Greenfield

**Deputy Art Director:** Yasia Williams-Leedham

**Picture Researchers:** Giulia Hetherington and Jennifer Veall

**Senior Production Manager:** Peter Hunt

# PHOTO CREDITS

Star maps by Wil Tirion, © Philip's

**Adam Block**/Mount Lemmon SkyCenter/University of Arizona 241. **Alamy** Alan Novelli 58; Chris Martin 174-175; Ian Dagnall 154-155; INTERFOTO 146; Jeff J Daly 19; Michael Kemp 25; Otto Mehes 44; Patrick Eden 54; Phil Wills 57; Pictorial Press Ltd. 17, 147, René Mattes/hemis.fr 182; RGB Ventures/SuperStock 52, 129; Timewatch Images 41; WorldPhotos 236; ZUMA Press, Inc. 231. **Bridgeman Art Library** Private Collection 185. **Caltech Archives** 268. **Corbis** 67, 81; Bettmann 18; Dennis di Cicco 161; EPA/NASA 107; Heritage Images 15; Jon Hicks 2; JPL-Caltech/Martin Benson 114; Luis Argerich/Stocktrek Images 70; NASA/JPL-Caltech/Michael Benson/Kinetikon Pictures 118; Roger Ressmeyer 47, 77, 137 below; Tony Hallas/Science Faction 221. **Dietmar Hager** (www.stargazer-observatory.com) 224. **ESO** 215, 253; Babak A Tafreshi/TWAN 22-23; C. Malin 167; Gianluca Lombardi 39; H. H. Heyer 4-5, 142-143, 254-255; J. Emerson/VISTA. Acknowledgment: Cambridge Astronomical Survey Unit 262; NASA's Goddard Space Flight Center/F. Reddy 230; T. Preibisch 218; VPHAS+ team 219; WFI (Optical), MPIfR/ESO/APEX/A.Weiss et al. (Submillimetre), NASA/CXC/CfA/R.Kraft et al. (X-ray) 265; Y.Beletsky 173. **European Space Agency** DLR/FU Berlin (G Neukum) 104; MPAe Lindau 163. **Fred Espenak** 199, Galaxy Picture library 53; Damian Peach 113, 133; Ed Grafton 141; Erwin van der Velden 97; John Rogers 125; Juan Carlos Casado 171; Peter Shah 343; Robin Scagell 42 below, 100, 225; Thierry Legault 79. **Getty Images** British Library/Robana 279; Buyenlarge 281; Denver Post 87; Eckhard Slawik/Science Photo Library 375; James Balog 158; Jamie Cooper/SSPL 48; Jay M. Pasachoff 207, 227, 237; Leemage 278; Mansell/Time & Life 40; Michael Rougier/Time & Life Pictures 232; Pat Gaines 260; Royal Society, London/Bridgeman Art Library 46 below; Science Museum/SSPL 276-277; SSPL 123; Stocktrek Images 191, 223; The Asahi Shimbun 153; The Bridgeman Art Library 242; UIG via Getty Images 177, 186; Walter Rawlings 159; William Radcliffe/Science Faction 68. **Marsyas** 13. **Martin Mobberley** 124. **Mary Evans Picture Library** ©Photo Researchers 169. **NOAO**/KPNO 56, **NASA** 101, 103, 105, 145, 197 right, 239, 66, 197 left, 200-201; ESA, M. Livio and the Hubble 20th Anniversary Team (STScI) 202-203; Apollo 15 82; Apollo 16 71; CXC/M.Weiss 248-249; CXC/SAO/F.Seward 247; Dr. R. Albrecht; ESA/ESO Space Telescope European Coordinating Facility 157; ESA and the Hubble Heritage Team (STScI/AURA) 250-251; ESA and the Planck Collaboration 272; ESA and The Hubble Heritage Team (STScI/AURA) 217; ESA, C.R. O'Dell (Vanderbilt University) and D. Thompson (Large Binocular Telescope Observatory) 238; ESA, CXC and the University of Potsdam, JPL-Caltech and STScI 258; ESA, J. Hester and A. Loll (Arizona State University) 246; ESA, K. Sahu and J. Anderson (STScI), H. Bond (STScI and Pennsylvania State University), M. Dominik (University of St. Andrews) 229; ESA, NASA, K. Sharon (Tel Aviv University) and E. Ofek (Caltech) 275; ESA, S. Beckwith (STScI) and The Hubble Heritage Team (STScI/AURA) 266; ESA, W. Keel (University of Alabama) and the Galaxy Zoo Team 63; ESA/CXC/University of Potsdam/JPL-Caltech/STScI 6-7; ESA/Herschel/PACS/L. Decin et al. 234; ESA/S Beckwith(STScI) and The HUDF Team 21; Goddard Space Flight Center 195; H. Ford (JHU), G. Illingworth (UCSC/LO), M.Clampin (STScI), G. Hartig (STScI), the ACS Science Team and ESA 267; Hinode, JAXA 61; JHU APL/CIW 95; JPL 99, 115, 120-121, 131 centre, 139; JPL/Space Science Institute 117, 131 right; JPL/USGS 74; JPL/USGS 138; JPL-Caltech 137 above; JPL-Caltech/IRAS/H. McCallon 205; JPL-Caltech/NOAO/AURA/NSF 256; JPL-Caltech/Space Science Institute 88-89, 130; JPL-Caltech/SSI/Cornell 127; JPL-Caltech/UCAL/MPS/DLR/IDA 151; NASA/CXC/PSU/S.Park & D.Burrows (X-ray), NASA/STScI/CfA/P.Challis (Optical) 245, NASA/CXC/SAO (X-ray), NASA/JPL-Caltech (Infrared), MPIA, Calar Alto, O.Krause et al. (Optical) 243; Neil Armstrong 85; NSSDC 75; SDO 211; SDO/AIA 178-179/Goddard Space Flight Center 192-193; SDO/Goddard Space Flight Observatory 178-179; The Hubble Heritage Team (STScI/AURA) 264; U.S. Geological Survey 131 left; National Radio Astronomy Observatory: Image courtesy of NRAO/AUI 269. **© Philip's** 26, 42; **Science Photo Library** 377; Royal Observatory Edinburgh 51; Royal Astronomical Society 135, 224; David Van Ravenswaay 91; Frank Zullo 165; Luke Dodd 235; Mark Garlick 207; National Optical Astronomy Observatories 213; Robert Gendler 261; Royal Astronomical Society 135, 224; Shaun Lowe, Alex Hubenov 183; Antlia 46; Filip Fuxa 13; Igor Kovalchuk 370-371; Marcel Clemens 92-93; Migel 152; Paulo Afonso 233; Suppakij1017 64-65; Suratchet Meeuwew 9; Viktor Klepnik 73; Wolfgang Kloehr 263. **SuperStock** JAGOPPN/BSIP 189; m-gucci 189; Volker Springel and the Virgo Consortium 271; iStock 16 left, 16 right;